Privatization for the Public Good?

Welfare Effects of Private Intervention in Latin America

Alberto Chong
Editor

Inter-American Development Bank

David Rockefeller Center for Latin American Studies
Harvard University

©2008   Inter-American Development Bank
        1300 New York Avenue, N.W.
        Washington, D.C. 20577

Co-published by
David Rockefeller Center for Latin American Studies
Harvard University
1730 Cambridge Street
Cambridge, MA 02138

Produced by the IDB Office of External Relations

To order this book, contact:
IDB Bookstore
Tel: (202) 623-1753
Fax: (202) 623-1709
E-mail: idb-books@iadb.org
www.iadb.org/pub

The views and opinions expressed in this publication are those of the authors and do not necessarily reflect the official position of the Inter-American Development Bank.

**Cataloging-in-Publication data provided by the
Inter-American Development Bank
Felipe Herrera Library**

Privatization for the public good?: welfare effects of private intervention in Latin America / Alberto Chong, editor.

      p. cm.
      Includes bibliographical references.
      ISBN: 978-1-59782-060-8

1. Privatization—Latin America—Social aspects. I. Chong, Alberto. II. Inter-American Development Bank. III. David Rockefeller Center for Latin American Studies.

HD4009 P84 2007   LCCN: 2007940419

Cover Design: Ultradesigns.com

Table of Contents

Foreword

According to many opinion polls there is widespread dissatisfaction with the privatization process in Latin America. This dissatisfaction has led some governments in the region to reverse some privatizations and return to a model of state ownership and operation. In at least some cases, however, these opinions are not supported by the empirical evidence available, and government reversal of privatizations may be the incorrect response to perceived or actual deficiencies.

This book provides a detailed microeconomic analysis of the impact of various individual privatizations in different countries in the region. Its central message is that in many cases, contrary to popular belief, society as a whole—and in particular the poor—have benefited from privatization. The book presents a careful analysis of the various mechanisms through which privatization has an impact on welfare, an analysis that by and large has been missing from the debate. For instance, because of the waterworks privatization in Argentina studied in this volume, many marginal neighborhoods that previously lacked service now enjoy water connections; as a result of improved water quality cases of diarrhea have decreased significantly, and reported cases are less severe. Since service delivery has improved, residents no longer have to pay for cistern trucks to deliver water to their homes or, worse yet, collect water with buckets from streams. As a result of better service and improved quality, consumers have saved both time and money, even if the prices charged by the privatized utility are higher than those charged by the public utility. Similarly, the book argues that to evaluate the Argentine electricity privatization it is necessary to consider the effect of improved electricity service on the quality of food consumed by households and, in turn, on children's health. Evidently, if the supply of electric power is erratic and subject to frequent interruptions refrigerators do not operate well. The book documents how, as power supply improved after privatization, refrigerators worked better and food did not spoil. As a result, there was a reduction in low-weight births and infant mortality.

Or consider the impact of privatizing the telephone service on the rural poor in Peru. Because the government required the private company to install public telephone booths in randomly selected villages it is possible to assess the impact of having access to telephone services and compare the differences between villages with and without service. Compared to residents in villages without telephone booths, beneficiaries in more fortunate towns with telephone service enjoyed measurable increases in income, especially income derived from nonagricultural activities; this helps to both increase income levels and reduce the overall variance of the rural poor's total income. While government requirements rather than the goodwill of the private company were responsible for these benefits, this only highlights the potential high returns of well-designed privatizations, as the public company was either unable or unwilling to provide this type of service in the past.

While it is not possible to generalize these conclusions to all cases analyzed, the general message of this volume is that privatizations can be beneficial to society at large, and that the charge that privatizations only benefit rich investors at the expense of poor consumers is not necessarily true in all cases. The research presented in this book is critical to the debate in the region. In particular, it points out that governments operating under the assumption that privatizations will always reduce welfare may in fact end up hurting the citizens they profess to protect. The evidence presented in this volume makes a strong case that privatization, when accompanied by appropriate regulation, works to the benefit of all, including the poor. With careful empirical evidence from six case studies, the book shifts the terms of the debate on privatization and brings to the forefront regulatory issues that should be at the center of the discussion. If private ownership accompanied by careful regulation can lead to gains for all, reverting to public ownership will most likely not solve regulatory problems, but will instead deprive the country's citizens of significant welfare gains. By presenting systematic empirical evidence to clarify a much-debated issue, and by carefully analyzing the arguments surrounding this topic, this book makes a very important contribution in an area critical for Latin America's development.

Santiago Levy
Vice President for Sectors and Knowledge
Inter-American Development Bank

Acknowledgments

This book was written with the support of the Latin American Research Network at the Inter-American Development Bank. Created in 1991, the aim of this research network is to leverage the Bank's research department's capabilities, improve the quality of research performed in the region, and contribute to the development policy agenda in Latin America and the Caribbean. Through a competitive bidding process, it provides grant funding to leading Latin American research centers to conduct studies on the economic and social issues of greatest concern to the region today. The network currently comprises nearly 300 research institutes all over the region and has proven to be an effective vehicle for financing quality research to enrich the public policy debate in Latin America and the Caribbean.

Many individuals provided comments and suggestions: Guillermo Calvo, Florencio López-de-Silanes, Eduardo Lora, Gustavo Márquez, Hugo Ñopo, Ugo Panizza, Vanessa Ríos, Victoria Rodríguez, and Luisa Zanforlin. In particular, Suzanne Duryea and Eliana La Ferrara, co-supervisors of the particular Research Network that coordinated the previous drafts of the chapters in this book and gave insightful comments and suggestions. The editor also thanks two anonymous reviewers and colleagues who participated in additional formal and informal discussions. Extremely valuable input was also provided in the production of this book by Rita Funaro and John Dunn Smith. Additional valuable support was provided by Norelis Betancourt, Rafael Cruz, Raquel Gómez, Luis Daniel Martínez, Michael Harrup, María Helena Melasecca, and Elisabeth Schmitt.

CHAPTER ONE

On the Social Effects of Privatizations in Latin America

Alberto Chong[1]

After the widespread crises of the 1980s, most Latin American countries embraced institutional and economic reforms intended to reduce fiscal deficits and inflation, liberalize economies, and modernize the state apparatus (Lora, 2007). Such reforms inevitably changed the relationship between citizens and the state. Before the reforms the state was seen as a social benefactor, serving as both a large-scale employer and a provider of a vast array of goods and services through active participation in markets. This view has changed substantially in recent decades.

While the economic and institutional effects of reform have been extensively analyzed, only recently has research focused on the consumer side, particularly the welfare effects of the reforms. This book constitutes a contribution to this line of research, presenting evidence on the effects of privatizations on several areas of consumer welfare. The research collected in this book analyzes the effects of the privatization of public services such as water, electricity, and telephones, in four Latin American countries.

Latin Americans have generally disapproved of privatizations since the initial stages of structural reforms in the early 1990s. Nevertheless, opinion poll approval ratings for privatization have substantially increased in recent years as the benefits of specific privatizations have become more

[1] I appreciate comments from Suzanne Duryea, Eliana La Ferrara, Gianmarco León, Eduardo Lora, Florencio López-de-Silanes, and Máximo Torero. Vanessa Ríos provided excellent research assistance.

apparent. According to this trend, a growing share of the population is able to enjoy the benefits of privatized enterprises, especially in improved access to and quality of basic services, which in turn improve access to a wide array of economic and social activities. Recent progress notwithstanding, approximately two out of every three Latin Americans take a negative view of privatizations. In the last round of polling by Latinobarómetro (2006), covering 17 countries, only 30 percent of Latin Americans said that they were "satisfied or very satisfied" with the results of the privatization of public services, "considering price and quality."

What shapes Latin Americans' opinions of privatization? Gaviria (2006) has shown that support for privatization is closely linked with wealth and the perception of social mobility. The richest quintile of the region's households are on average 8 percent more likely to approve of privatizations than the poorest quintile. Also, regardless of their income level, households whose members perceive that they have experienced (or may experience in the future) social mobility are significantly more likely to approve of privatizations.

Other explanations for popular discontent—and perceptions that privatization has exacerbated social exclusion—suggest that this discontent stems from a widespread belief that privatizations have given private investors, who are seen as members of the economic elite, control over assets considered important for the country (Birdsall and Nellis, 2002). These feelings became especially apparent during the privatization of telecommunications in Mexico and Peru.

Popular approval of privatizations is further diminished by the absence of a political consensus on which activities should be under government control. According to a 1998 survey conducted by the *Wall Street Journal Americas* in 14 Latin American countries, on average 31 percent of those interviewed thought that airlines should be under government control, with 26 percent supporting government management of television stations and 61 percent believing that water services should be provided by the government.

It should also be noted that public perceptions of who benefits from privatizations, at least in the short term, have some basis in fact. Those who approved of privatizations in the *Wall Street Journal Americas* poll were also those who identified themselves as belonging to the political right, being in the higher deciles of the income distribution, and actively participating in—as well as being the first beneficiaries of—economic improvements in their countries. Though some may benefit more than others in the short run, there is no definitive answer to questions such as whether privatizations ultimately benefit groups that are already worse off in favor of wealthier interests.

Most research on the effects of privatizations has evaluated the efficiency or productivity gains of private over public management. These issues, however, hold little immediate interest for the public at large, who are probably much more concerned about direct welfare effects. At the household level, these include basic features of welfare, such as access to public services and their price. Additional direct effects on households include job losses, a diminished ability to unionize, cuts in social benefits and a greater sense of job insecurity. Wider concerns include corruption scandals arising from some privatization processes and the enrichment of particular individuals or companies as a result of privatization.

While all of these factors are important in forming public opinion and have serious economic consequences, they do not necessarily lead to the conclusion that privatization has negative social or distributive effects. With these concerns in mind, the current volume explores whether widespread popular disenchantment with the privatization process in Latin America is supported by evidence such as performance and employment indicators. Given this broad objective, this volume is an important contribution to the growing literature on privatizations in Latin America and other parts of the world.

The book studies the results of the privatization process on important outcome variables, such as efficiency, employment, and the effect on consumers' income. First, in terms of efficiency the evidence presented demonstrates that most companies for which there is data available showed a marked improvement in productivity or efficiency measures after they were privatized; the one notable exception occurs in Guayaquil, Ecuador.

Second, the book examines how privatizations affect the welfare of consumers. The book explores the effects of higher tariffs for services and the expansion of coverage on consumers of different privatized services in different countries. The evidence on privatizations in different South American countries reveals that, contrary to popular belief, the net effect of privatization on most income brackets was mostly positive; this resulted largely from expanded coverage.

Finally, the book explores the effect of privatizations on inequality and poverty in the countries covered. The net effect seems to be of an improvement in poverty and a reduction in inequality in most cases. This includes, for example, a reduction in common diseases among the lowest income brackets once there is an improvement in the quality of water provided by a privatized company.

The general conclusion of this volume is that privatizations can be socially very beneficial, but delivering these benefits to the poor requires the power of government to regulate the companies. The publication of such findings, however important, represents only a first step. Academic research must be followed by the hard work of changing public perceptions, which generally give greater weight to the "collective perception" of tragedy (e.g., job losses) than to the benefits that individual consumers observe.

This book thus represents a first attempt to provide systematic econometric evidence on whether certain privatization episodes led to an increase in social exclusion in the countries that undertook reforms.

Extent of Privatization and Its Effect on Efficiency

It is generally agreed that privatizations brought significant gains in productivity and efficiency.[2] This is notable because Latin America was a pioneer in promoting private participation in infrastructure projects. Between 1990 and 2003, about half of the total private-sector investment of US$786 billion in developing countries was directed to the region. The role of different sectors in the privatization process can roughly be gauged by their respective shares of total privatization revenue. In Latin America, 75 percent of that revenue has derived from the public service and infrastructure sectors, 11 percent from the financial sector, and the rest from the oil, fuel and manufacturing sectors.

---

[2] See Earle and Gehlbach (2003) for an example.

Nonetheless, the extent of privatization has varied across industries, and in no country have all state-owned enterprises (SOEs) been privatized. Most Latin American countries have privatized their telecommunications, electricity, fuel and, to a lesser extent, water and sanitation services. In contrast, privatization of railway companies, airlines, airports and expressways has been less widespread. Privatization of the financial and industrial sectors has not been important because private participation in these sectors was already extensive. Most countries have maintained the presence of at least one official bank and retained government control over companies related to natural resources such as oil, natural gas, and copper. Even Chile, one of the countries that most aggressively embraced the privatization of state-owned enterprises, has maintained official control of companies in key sectors such as copper, oil, banking, the postal system, railways and ports.

Within the overall trend toward privatization, its extent differed greatly across countries. As shown in Figure 1.1, for example, some countries with large state sectors such as Costa Rica, Ecuador and Uruguay privatized only a few companies, while other countries, including Argentina, Bolivia, Panama and Peru, sold off state companies for values of more than 10 percent of GDP. Uruguay was the only country that did not privatize companies in the electricity, oil and telecommunications sectors, perhaps because the privatizations were explicitly subject to a popular referendum, a mechanism used in no other country in the region. At the other extreme, Argentina underwent a privatization process that affected practically all infrastructure sectors, as well other sectors where the state was involved; notable exceptions involve some provincial health companies as well as some national and provincial banks.

Apart from the need to raise fiscal revenues, state companies have been transferred to the private sector primarily to achieve greater efficiency. Before privatization, Latin American SOEs had largely displayed failings common to firms managed according to political criteria: decisions were made in regard to employment, investment, location or innovation that proved detrimental to profitability and efficiency, thus producing fiscal deficits and undermining institutional frameworks. Evaluating the microeconomic effects of privatization in several Latin American countries,

Figure 1.1 **Privatization Revenues in Latin America and the Caribbean, 1990–2000** (Percentage of GDP, 1999 dollars)

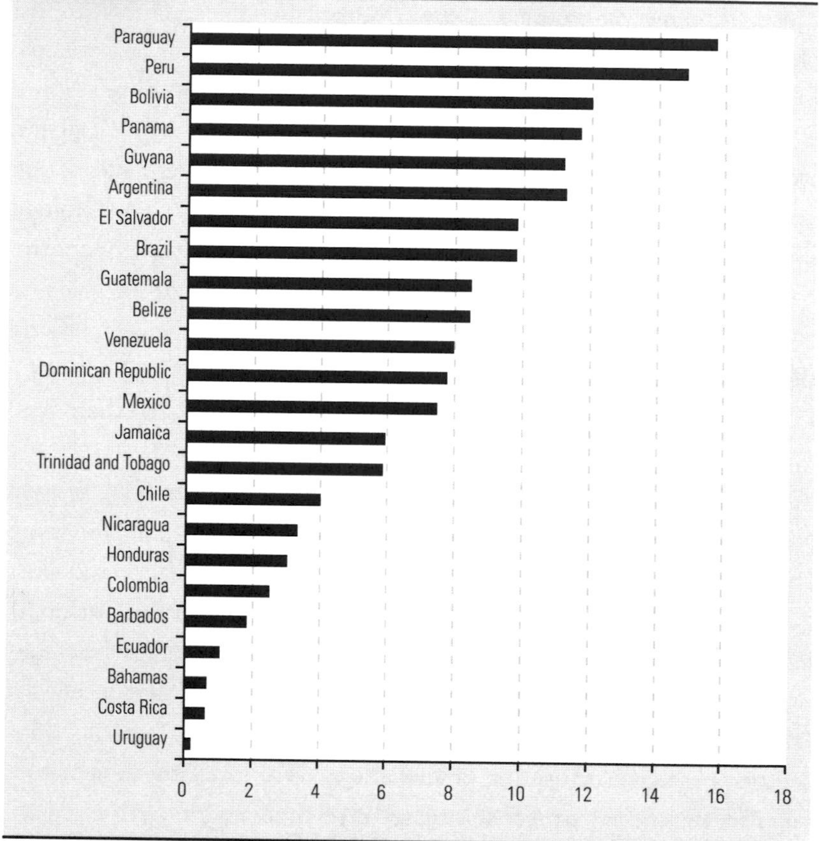

Source: Chong and López-de-Silanes (2005).

Chong and López-de-Silanes (2005) found that privatization considerably improved companies' profitability and efficiency. Typically, after privatization, companies increased their net income to sales ratio by 14 percentage points, mainly through improved efficiency, as unit costs dropped by an average of 16 percent. Other indicators yield similar results.[3] Figure 1.2 summarizes the study's main findings related to profitability.

---

[3] For example, the sales to assets ratio increased on average 26 percent, and the sales to employee indicator rose notably as well. In Chile and Mexico, the two most outstanding cases, the sales per employee ratio doubled in privatized companies and in certain com-

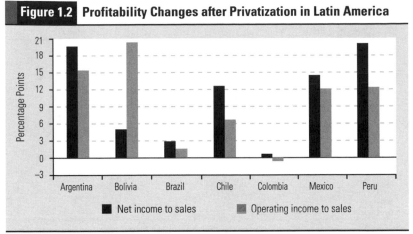

Figure 1.2 Profitability Changes after Privatization in Latin America

Source: Chong and López-de-Silanes (2005).
Note: The components of the variables are defined as follows: net income is equal to operating income minus interest expenses and net taxes paid, as well as the cost of any extraordinary items; operating income is equal to sales minus operating expenses, minus cost of sales, and minus depreciation; and sales are equal to the total value of products and services sold, nationally and internationally, minus sales returns and discounts. For Bolivia, the net income to sales ratio is not available.

Increasing productivity, however, is generally perceived to have come at the cost of labor force reductions and social benefit cuts. In particular, it is commonly held that most workers dismissed from public enterprises are forced to enter the informal sector, thereby losing a stable source of income and access to social benefits. This view is not without a basis in fact: since public companies have often been used to create employment for political reasons, short-run job reduction was necessary to make these companies viable as part of the privatization process.

The magnitude of job losses due to privatization in six Latin American countries is shown in Figure 1.3. While industry-adjusted job losses in Chile averaged only about 5 percent, in Peru and Argentina the industry-adjusted average of job reduction in privatized SOEs was

---

panies those increases were several times larger. These results might at first glance seem to stem from simply lowering costs and reducing the workforce. In fact, however, in the wake of privatization these indicators improved even as company production substantially increased. Mexico and Colombia registered the greatest average gains of 68 percent and 59 percent, respectively. Brazil, which trailed the other countries in the study, nonetheless increased production by an impressive 17 percent.

Figure 1.3 Percentage Changes in Employment after Privatization in Latin America

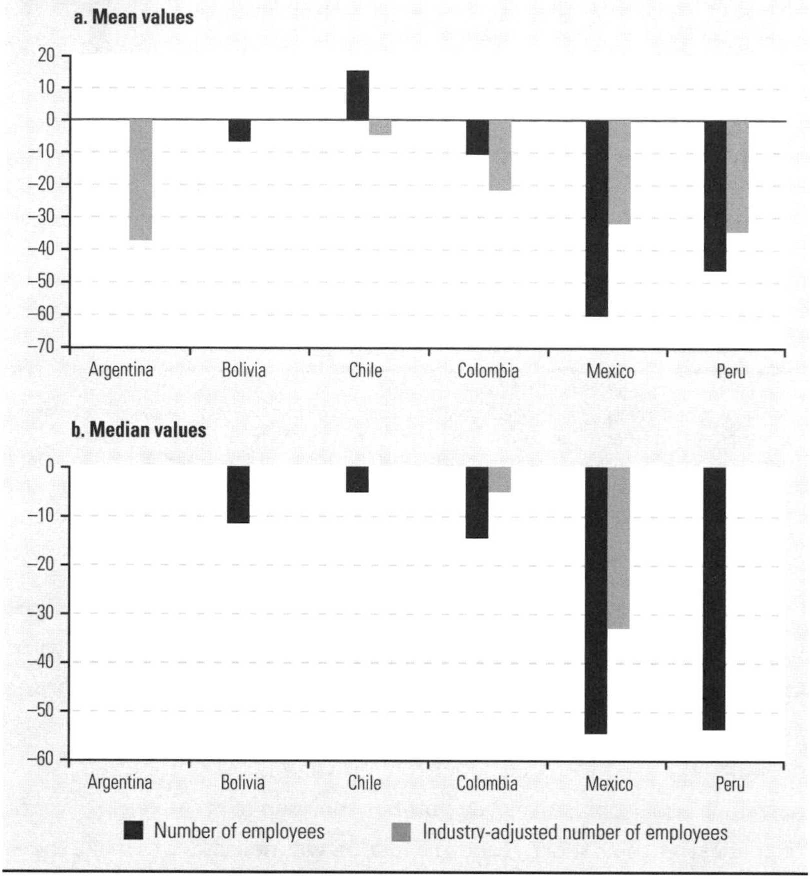

a. Mean values

b. Median values

■ Number of employees   ■ Industry-adjusted number of employees

Source: Chong and López-de-Silanes (2005).
Note: The number of employees corresponds to the total number of workers (paid and unpaid) who depend directly on the company. The industry-adjusted number of employees is computed by augmenting the preprivatization number by the difference between the cumulative growth rate of the number of employees of the firm and the cumulative growth rate of the number of employees of the control group in the postprivatization period relative to the average number of employees before privatization. For Argentina, the mean number of employees is not available; for Chile and Peru, the median industry-adjusted information is not available; for Bolivia, the industry-adjusted information is not available.

around 35 percent. However, the effect of privatizations on unemployment in Colombia seems to have been very modest, at least in the electricity sector, where most privatizations took place (Chong and López-de-Silanes, 2005).

These findings shed light on some important short-term consequences of privatization episodes, but do not allow us to assess the medium- to long-run effects of these reforms. In fact, in the medium term, many firms ended up rehiring workers who had initially been fired during the privatization processes—once it became that clear that the "wrong" workers had been dismissed. As shown in Figure 1.4, privatizations in Latin America have offered a prime example of this so-called "adverse selection" problem. Some workers who were let go did in fact end up in the informal sector (Chong, López-de-Silanes and Torero, 2007), indicating that social exclusion as a result of privatization layoffs did occur, but this problem was in some cases mitigated by the rehiring of those workers. According to the data in Figure 1.4, about 20 percent of previously dismissed workers in Latin America were rehired by the same firms—the highest percentage in any region.

Whether the efficiency gains from privatization were driven by productivity-enhancing investment, or by reductions in jobs and social benefits, remains unclear. Although in some large countries many workers were dismissed as a result of privatization, in other countries privatized firms actually created a significant number of new jobs in the medium term.

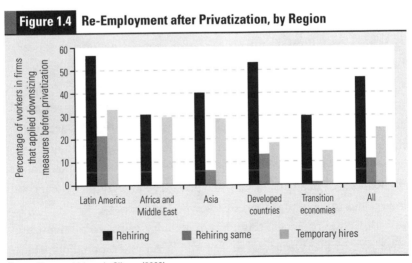

Figure 1.4 **Re-Employment after Privatization, by Region**

Source: Chong and López-de-Silanes (2003).
Note: Original list based on 1,500 firms. Sample reflects the 225 firms that applied downsizing measures before privatization.

Further Labor Effects of Privatizations

Given the topic of this book, it is important to give some more background on the labor effects of privatizations. Critics of privatization maintain that employment reductions are both the primary means of driving up productivity and the major cause for the exclusion of low-skilled and elderly workers from the formal labor market. While the limited evidence available suggests that labor cost reductions contribute to profitability gains after privatization, these savings do not explain the bulk of increased profitability (La Porta and López-de-Silanes, 1999). Moreover, job reductions are not the only means of increasing labor productivity and, even when they occur, they may be accompanied by other cost-cutting measures such as lower wages and benefits.

Productivity, though, cannot be viewed in isolation. Taking into account other labor indicators and using the World Bank's Productivity and the Investment Climate Survey (2006), which provides considerable detail on the microeconomic and structural dimensions of 75 countries' business environments, Chong and León (2007) find mixed evidence on the benefits of privatization. Managers in privatized firms earn significantly higher wages than their counterparts in either state-owned firms or firms that have always been private, while the wages of lower-skilled workers in privatized firms do not differ significantly from similar workers in private or state-owned firms. On the other hand, working conditions appear to have significantly deteriorated in the transition from public to private ownership. In addition to labor deregulation throughout the region, there has been a clear trend of reduction in non-wage labor costs, especially social benefits. In other words, privatized and private firms seem to be favoring temporary workers over permanent contracts and employing more low-skilled workers. Under these circumstances, workers' ability to organize has been diminished, and privatized firms display significantly lower unionization rates than state-owned enterprises.

In some instances, though, productivity gains from privatization have led to higher wages and other forms of remuneration. For instance, in Mexico, La Porta and López-de-Silanes (1999) find that wages in a broad sample of privatized companies increased an average of 76 percent from

1983 to 1994, well above the rest of the economy. Even more surprising, wages increased substantially more for blue-collar workers than for office staff (122 percent compared with 77 percent in the 1983–94 period). Workers in many privatized companies additionally benefited from ownership participation programs introduced to increase worker interest in privatization. In Colombia, Pombo and Ramírez (2005) find that average wages in privatized manufacturing firms increased by 25 percent after privatization. As in other countries, however, it appears that other labor conditions have deteriorated and the influence of labor unions has been eroded (Chong and López-de-Silanes, 2005).

A further concern is what happened to those workers who were laid off during the restructuring process either before or after privatization. This segment of the population, usually drawn from the groups that are most vulnerable to economic shocks of any kind, arguably faces the greatest risk of social exclusion due to privatization. In fact, one of the leading concerns surrounding privatization has been that laid-off workers may be unable to obtain comparable jobs in the private sector because of age, low skills, or the accumulation of industry-specific human capital. Although data limitations have made it nearly impossible to seriously address this issue on a large scale in most countries, Chong, López-de-Silanes and Torero (2007) have analyzed the conditions of laid-off workers in Peru both before and 10 years after privatization. They find that, even though laid-off workers were given a compensation package, the average worker suffers a significant initial hit after being fired, which validates concerns regarding the impact of privatization on inequality. On the other hand, the evidence provided by these researchers shows that workers' wages and benefits eventually recover up to a similar level to those workers in the private sector in the corresponding industry at the country level.

Perhaps more surprising is that "stayers" in Peru command higher wages and benefits than comparable workers who had been fired because of privatization or who had always been in the private sector. Workers in former SOEs were apparently able to extract more than other workers due to firm market power, union power or favorable terms of a collective contract that remains in effect. This result helps to explain why the compensation of workers who lost their jobs because of privatization reverts to the mean of their private sector industry. The same study reveals that the typical worker consumes the compensation package within the first two years, usually through investment in the home or through the creation of a self-run business. Unfortunately, however, the average new business fails by the end of that period and workers move on to activities related to their previous employment. Although limited to the case of Peru, these results help to qualify the belief that workers lose in the long term after privatization and that workers who lose their jobs are condemned to unemployment or poverty. Other relevant studies which focus on wages and employment are Tansel (1998), who uses retrospective data for Turkey; Galiani and Sturzenegger (2005), who focus on one firm in Argentina; Haskel and Syzmanski (1992), who study the United Kingdom; Brown, Earle and Vakhitov (2006), who study Ukraine; Chong, Rodríguez and Torero (2007) for Mexico; and Chong and López-de-Silanes (2005), who show that adverse selection plagues privatization retrenchment programs, which casts doubt on the negative impact on employment. In general, it appears that workers laid off as a result of privatization experience a significant setback at first, then recover and converge towards a situation in which wages and benefits reach levels similar to those in the private sector.

The Role of Discrimination

An important component of social exclusion relates to the *criteria* used in distributing gains and losses from reforms. Studies of racial discrimination in labor markets have focused on documenting earnings differentials between females and males, or indigenous and non-indigenous people, or Afro-descendants and whites. Comparisons of hourly labor earnings suggest the existence of conspicuous gaps: depending on the estimates used, non-indigenous workers earn between 80 percent and 140 percent more than their indigenous counterparts (Inter-American Development Bank, 2007). However, non-indigenous workers exhibit human capital characteristics that are, on average, more desirable than those of indigenous workers. The most notorious of these characteristics has been education (schooling), but there have also been differences in labor market experience and field of specialization. Therefore, to attribute the whole earnings gap to the existence of labor market discrimination in pay would be misleading. At least a part of that gap can be traced to differences in observable human capital characteristics that the labor market rewards and, hence, has nothing to do with discrimination. Econometric techniques have helped to identify, to some degree, the magnitude of this component. In terms of racial earnings gaps, these differences in human capital characteristics account for more than one-half of the documented earnings gaps.

In many industrialized as well as developing countries, one of the prominent features of discrimination in labor markets is that it occurs on the basis of race or ethnicity. A study of the privatization process in Peru examined the ethnicity of laid-off workers and found descriptive evidence of a bias against the non-white population. However, a statistical analysis considering a number of other factors refutes the discrimination hypothesis, at least when it comes to dismissing workers. Instead, the human capital endowments of individuals determined whether they were fired or not. Of course, discrimination may have been important in determining whether these individuals had the type of education or health care they needed to be competitive workers, but it was not the determining factor when they were laid off following privatization (Ñopo, Saavedra and Torero, 2007; Chong, Nakasone and Torero, 2007).

According to conventional wisdom, Latin America is a highly discriminatory society. However, the results of this study raise the question of who is discriminated against? The region's quintessential opinion survey, Latinobarómetro, finds that most people in the region think there is some sort of discrimination. However, most Latin Americans do not believe that it operates against groups who are traditionally discriminated against (indigenous peoples, Afro-descendants and women, to cite the most prominent historical examples). Instead, they believe the poor are the ones who suffer the most from unequal treatment. After the poor, Latin Americans believe that the uneducated, the elderly and those who lack proper social connections are those who suffer the most from discrimination. Interestingly, nearly a quarter of respondents said they did not feel discriminated against at all. These perceptions of discrimination are counterintuitive and are summarized in Figure 1.5. They point towards the

Figure 1.5 Reasons for Unequal Treatment That Most Affect Population, 2006

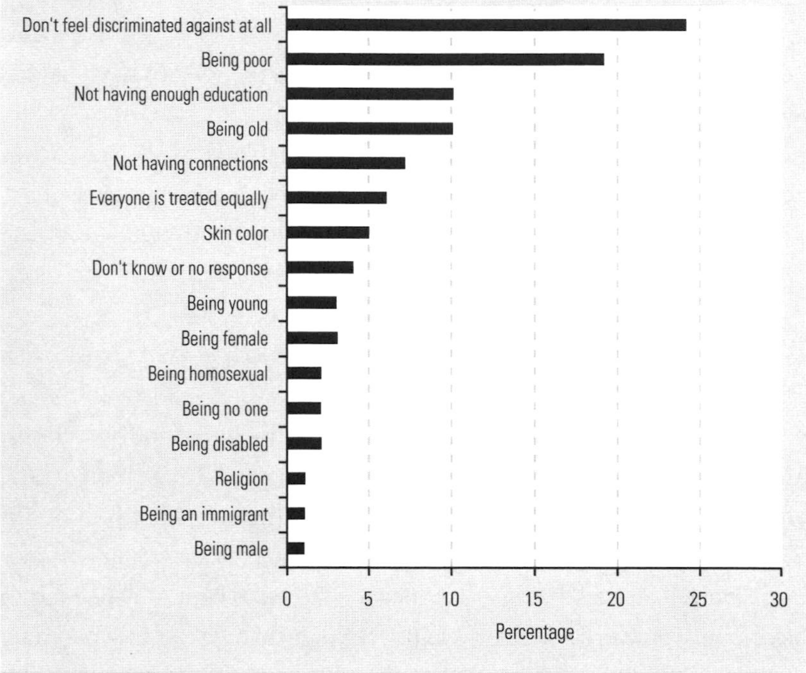

Source: Latinobarómetro (2006).
Note: Figure reflects responses to the question "Of all the reasons for which people are not treated equally, which one affects you most?"

existence of some sort of discrimination derived on the basis of economic reasons, rather than biological or sociological factors.

The evidence of discrimination (or, more precisely, earnings gaps that cannot be explained by differences in productive characteristics of individuals) that this type of study has found is conspicuously smaller than a simple comparison of earnings would suggest (Ñopo, Saavedra and Torero, 2007). Nonetheless, these studies have been criticized, particularly for their failure to truly identify discriminatory behaviors due to the presence of "unobservable characteristics." That is, the human capital characteristics these studies can typically analyze are only those that are easily observable (schooling, labor market experience, field of specialization, sector choice, etc.), but there are other unobservable characteristics—including entrepreneurship attitudes, motivation, work ethic, commitment and assertiveness—that are intrinsically harder to measure. These characteristics are typically not captured in a survey (and in that sense, are not "observed"), but they may be observable by an employer, or more generally, the relevant actors in the labor market.

Recent research on the so-called discriminatory practices in Latin American labor markets attempts to separate the effect of observable from unobservable characteristics.[4] The results represent a giant step towards the goal of understanding discrimination and its channels, using tools that emphasize efforts to "observe the unobservables." Interestingly, many of the results obtained from controlled experimental settings seem to contradict the idea that Latin Americans act discriminatorily nowadays. The evidence found points towards the existence of stereotyping that vanishes when information is revealed.[5] To some extent, there is also evidence that some sort of self-discrimination partially explains the discriminatory outcomes. Women asking for wages between 6 percent and 9 percent lower than those asked for by their male counterparts is a case in point. Both stereotyping and self-discrimination are behaviors that may simply reflect the substantial differences in endowments of agents in the market. Under these circumstances, labor markets simply operate as resonance boxes that amplify differences that exist in other spheres.

Clearly, what most Latin Americans observe in their daily activities are substantial differences in human, physical, financial and social assets that are associated with gender, racial, ethnic and class distinctions. However, tracing the link between potential discriminatory practices and these outcomes is methodologically challenging, and it seems that the public discourse is not always informed in this sense.

---

[4] Examples of this research are Núñez and Gutiérrez (2004), Bravo, Sanhueza and Urzúa (2006), Bertrand and Mullainathan (2004), Moreno et al. (2004), Cárdenas et al. (2006), Castillo, Petrie and Torero (2007), Elías, Elías and Ronconi (2007), and Gandelman, Gandelman and Rothschild (2007).

[5] For instance, when choosing partners in an experimental investment game that depended not only on the individual's decisions but also on the decisions of their peers, people showed a preference for women, tall and white-looking people. However, when given information on the past performance of other players, the information previously used to stereotype does not seem to matter any more.

Distributive and Welfare Effects

In addition to the effects on labor markets, privatizations have important distributive and welfare effects through their impact on consumption. Typically, privatizations of public services bring important increases in *tariffs*, which affect low-income groups more strongly. However, the effect of expanding coverage of services operates in the opposite direction, as relatively richer households typically already have access to public utilities. Consequently the net effect on poverty and inequality is a combination of both factors.

A comparative study for Argentina, Bolivia and Nicaragua estimated these effects with very revealing conclusions (McKenzie and Mookherjee, 2003). In Argentina the effects were positive for all deciles in electricity and water privatizations. In electricity the benefits for the poorest 30 percent of the population represented 2–3 percent of income. In Bolivia, telephone privatization benefited above all the middle class, where expansion of coverage was most evident. In deciles 5 to 7, the net benefits were equivalent to 5 or 6 percent of income. The privatization of water in the La Paz and El Alto conurbation produced benefits of the same magnitude for the lowest deciles due to increased access, although the failed water privatization in Cochabamba may have produced negative effects for the poorest due to the tariff increase. In electricity, Bolivian privatization had positive effects for all deciles, except the richest. Finally, in Nicaragua electricity privatization produced small gains in deciles 2 to 6 and small losses in others.

The final effects on inequality and poverty seem to be favorable for most privatizations, although modest. The magnitudes of the estimates imply reductions of around 0.01 or less in the Gini coefficient and two percentage points in the poverty incidence rate. With one single exception among the cases considered, the popular perception that privatizations increase poverty or inequality is not valid. However, these studies (Birdsall and Nellis, 2005) offer only a partial overview of the social benefits of privatization and may be criticized for not accurately measuring the effects of improvements on the quality and availability of services and for shortcomings in the data employed (Saavedra, 2004).

The studies collected in this volume offer some evidence that there are benefits in addition to those measured by previous studies. For example, important benefits were found for health, time management and the employment possibilities of beneficiaries.

Two of the studies in this volume relate to Argentina, where most of the population actually has a negative opinion of privatizations. Galiani et al. (Chapter 2) analyze the effect of the expansion of waterworks by a privatized firm, which expanded coverage to urban marginal shanty-towns, where there was no service previously (although coverage is still not complete). The partial coverage of the water service allowed the researchers to assemble a panel data set that selected a control group from those households that had not benefited from the water service expansion. The results show that in the area of Buenos Aires access to piped water had significant effects on the reduction of diarrhea episodes among children. Also, when they were present, diarrhea episodes were shorter and less severe than those experienced by children from households that had not benefited from the expansion of the water services. Similarly, the improved provision of water produced substantial savings of money and time, because families did not have to spend time looking for other sources of water provision. An interesting and unexpected additional finding was that even households that had illegal connections before privatization received gains in health and time because, although the water might have been free, it was of very low quality. These results reinforce those from a previous study by Galiani, Gertler and Schargrodsky (2005), which used data from municipalities with privatized water services (about 30 percent of the country's municipalities) to show the effect of the private provision of water services on child mortality in Argentina. Their earlier results show that child mortality fell between 5 to 7 percent in areas that privatized their water services, and this effect was largest in poorest areas.

Also for the Argentinean case, González-Eiras and Rossi (Chapter 3) address the effects of the expansions of the privatized electricity networks on health outcomes. Their study adopts a novel approach to trace a link between electricity provision, durable goods ownership and nutritional content of the food consumed by household members. Their goal is to measure the effect that good electricity service (measured by access to and continuity of service) can have on the quality of food consumed in households, and how this can affect children's health. The underlying hypothesis is that in areas where the electricity supply is not stable and continuous, refrigerators do not work well, and this in turn affects the type or quality of food consumed by individuals. In their study, based on information collected at the household level, they find suggestive evidence that electricity privatization reduced the frequency of some public health problems usually caused by food poisoning and nutrition. In provinces where electricity distribution was privatized, the frequency of low birth weights decreased in the range of 3.7 percent to 5.8 percent relative to provinces with public networks, though results are not robust to accounting for the potential problem of serial correlation in the data. Similar results are found when evaluating the continuity and frequency of electricity provision on child mortality.

In their study of Colombia, Barrera-Osorio and Olivera (Chapter 4) also analyze the effects of water service privatizations on access to the service, its quality and price, and some associated health outcomes. In the absence of panel data at the household level, the researchers used two different household surveys: the Encuesta de Calidad de Vida and the Demographic and Health Surveys (DHS). They constructed a panel made up of municipalities where they were able to observe a group that actually privatized the water services and a control group where water services remained under the control of local authorities. Overall the results showed a positive effect of privatized water services on welfare indicators such as the number of children with diarrhea, which in rural areas shows a reduction of around 11 percentage points. Urban municipalities with private water provision saw a significant increase in the coverage of the network, and the water provided also had higher quality. When the quality of water is measured as the frequency of the service, it is higher in nonprivatized urban municipalities, although the second quintile presents an improvement in the frequency of the service compared to the fourth and fifth quintiles. There was also a redistribution effect toward poorer areas in the frequency of the service. While richer areas saw a significant reduction in the frequency of service, it increased in poorer areas.

Regarding health outcomes, improvements in the frequency and quality of water services in privatized municipalities significantly increased the weight for height index of children, which is taken as a proxy of health status. Despite the positive impact observed on health, the users saw a great increase of the prices of the water services, which may outweigh the other improvements in the quality and extension of the service. Given the information available, it was not possible for the researchers to disentangle the effect of the private enterprises and the cross-subsidies that were eliminated once the public network was divided into several local companies.

In the case of Peru, Chong, Galdo and Torero (Chapter 6) find that privatization of the telephone service brought important benefits for the rural poor. The government required the private company, Telefónica del Perú, to install public telephone cabins in some villages selected at random. Taking advantage of this "quasi-natural experiment," the study evaluates the effects of this decision on the beneficiaries, concluding that there were tangible improvements in income, especially non-agricultural income, which is crucial for stabilizing the income of the rural poor. In this way, total per capita non-farm income among the population who is treated ("existence of public telephone installed by the privatized firm") would be around 28 percent lower otherwise. In this case there were clear effects acting against social exclusion, reinforcing the mechanisms through which campesinos could generate a more stable source of income out of the agricultural work. Although these benefits were not the result of spontaneous initiatives on behalf of the company, but rather of government requirements, the results suggest that in circumstances where public companies may not have the means of extending coverage or improving the quality of the service, a combination of privatization and adequate government regulation may channel the benefits towards less privileged social groups.

Another study for Peru suggests that electricity privatization has substantially benefited peasants and poor rural workers (Alcázar, Nakasone and Torero, Chapter 7). This was established by comparing areas where electricity distribution was privatized with areas where control remained with the public sector. According to the study, the quality of service improved substantially in the privatized areas, which had an impact on the time management of users. More continuous and reliable electricity service enables users to spend less time on agricultural work and more on nonagricultural or leisure activities, with positive effects on income and welfare.

In addition to the material benefits that this can generate, a similar expansion of public services could help create a sense of inclusion in society among previously marginalized groups, and ultimately translate into a higher quality of life for Latin Americans.

The Challenge for Governments: Regulation

As pointed out by Chong and López-de-Silanes (2005), there are two main instances in which regulation should be carefully analyzed in conjunction with privatization: industries characterized as natural monopolies or in which oligopolistic market structures exist, and industries in which the government owns most of the assets even if no individual firm has substantial market power. Sectors with a heavy state presence tend to be protected by a web of regulations whose original intent was to cut SOE losses and reduce fiscal deficits. In some of these cases, the regulatory effort needed can be better understood as deregulation to get rid of protective structures that shield companies from competition and allow privatized firms to make extraordinary gains at the expense of consumers. Competition and regulation should be carefully considered as part of the aftermath of the privatization process. Adequate regulation can produce efficiency improvements that benefit both consumers and producers. In cases of sectors with oligopolistic power, the regulation effort needs to be complemented with a new package of rules and disclosures to enhance supervision and reduce abuse of market power.

There are two ways in which credible regulation complements privatization. At the most basic level, product market competition provides a tool to weed out the least efficient firms. This process may take too long, or not work at all, if regulation inhibits new entry or makes exit costly. Wallsten's (2002) analysis of the effects of telecommunications' privatization and regulation in Latin America and Africa shows that competition from mobile operators and privatization combined with the existence of a separate regulator are significantly associated with increases in labor efficiency, mainlines per capita and connection capacity. The implication is that privatization of oligopolistic industries with concurrent reforms may improve welfare, especially as it improves service. An additional bonus is that countries in which a regulatory agency existed prior to privatization were able to fetch higher privatization prices and thus provide governments with greater revenue that could be used for investment and social programs.

Secondly, adequate regulation may also complement privatization by raising the cost of political intervention. Whereas an inefficient monopoly can squander its rents without endangering its existence, an inefficient firm in a competitive industry would need a subsidy to stay afloat. The introduction of competition forces politicians to have to pay firms directly to engage in politically motivated actions whereas before the costs of these measures were absorbed by an SOE that did not have to worry about market performance. In fact, competition is often restricted precisely because it raises the costs of political influence.[6] As long as a suitable regulatory framework is in place at or before the time of privatization, both consumers and the government should benefit from the process, and it is therefore not surprising that governments that lack regulatory capabilities at the time of privatization and want to maximize sales prices have often postponed full and clear regulation. Trying to establish an adequate regulatory scheme after privatization may be problematic from a political economy perspective. Since the agency in charge of enforcing and regulating the contracts is often the same or subordinated to the agency that carried out the privatization, there is an incentive for lax enforcement to avoid exposing past mistakes. For numerous concessions in infrastructure projects, the private sector was able to bargain and maintain protective regulation after privatization because of the threat of bankruptcy, withdrawal, or desertion of future investment commitments.

---

[6] Colombia and Mexico provide good examples of adequate deregulatory policy actions that, when coupled with privatization, can be used as a lever to transform the economic landscape and reduce political interference in the economy in the early 1990s (Chong and López-de-Silanes, 2005).

All of these impact the reputation and credibility of privatizing politicians. In the last 15 years, concession contracts in developing countries have often led to renegotiations. In Latin America and the Caribbean, 40 percent of all concession contracts were renegotiated just over 2.2 years after they were signed. These opportunistic renegotiations of concessions are common because of what Engel et al. (2003) call a "privatize now, regulate later" approach. Cost overruns in concessions and unclear rules governing contingencies provide private owners with the opportunity to extract economic rents from the government. Finally, attempting to substantially alter the regulatory framework after the sale may also prove difficult as new constituencies against regulation are created at the time of privatization. Shareholders and managers of privatized SOEs are joined by workers and even consumers who could benefit from the protective regulatory status of firms (Chong and López-de-Silanes, 2005).

Adequate regulation can make privatization work—for governments and for consumers. Assuring that privatized firms operate in a competitive environment at the service of their clients rather than their stockholders is key to achieving the goal of a more efficient, productive and equitable economy. It may also help change the face of privatization in the eyes of public opinion.

Conclusions

While privatizations can be socially beneficial, delivering these benefits to the poor requires the power of government to adequately regulate the companies. This does not mean that Latin Americans are going to come out in favor of privatization, however, even if governments do things well.

A relevant factor in shaping public opinion is the tendency to group together simultaneous events or trends and consider them the common cause of short-term changes. In the economic policy area, the "Washington Consensus" or the "neoliberal economic model" are the common denominator under which a set of policies are grouped whose separate effects are difficult for an observer to disentangle. An objection to one outcome can lead to a rejection of all policies that observer correctly or erroneously associates with it.

Finally, there is a tension between the historical inefficiency of several state companies in Latin America and the ideological tenets of those who believe that basic services such as water provision should not be handed over to the private sector. While this volume does not take a stand on individual ideological positions, it is nonetheless important to inform the policy debate so that it is grounded on rigorous empirical analysis of the effects of reforms, as opposed to qualitative and descriptive evidence made of correlations.

CHAPTER TWO

Water Expansion in Shantytowns: Health and Savings

Sebastián Galiani, Martín González-Rozada and Ernesto Schargrodsky

E ven though water is one of the most crucial elements for life, more than 20 percent of the world's population does not have access to safe drinking water (WHO and UNICEF, 2004). Those who are not connected to the water system often resort to purchasing water from independent providers, and those who cannot afford it consume unsafe polluted water. This lack of access to safe drinking water is a serious threat to the health of disadvantaged populations, especially children.

At the 2000 Millennium Summit, the member countries of the United Nations unanimously agreed on a set of eight goals to reduce poverty by 2015, including reducing child mortality by two-thirds and halving the number of households that do not have access to safe water. These two goals are interrelated because clean water is critical to containing the spread of infectious and parasitic diseases. Indeed, each year more than 3 million children die from preventable water-related diseases (World Bank, 2002), and other studies have found that access to safe water is associated with better child health.[1] Galiani et al. (2005) document large improvements in access to water and water quality as a result of the privatization of water companies in Argentina during the 1990s. In addition, they find that privatization led to a reduction in child mortality in poor and extremely poor regions. In spite of these improvements, however, a large fraction of the Argentine population, mainly

---

[1] See Merrick (1985), Behrman and Wolfe (1987), Esrey et al. (1991), Lavy et al. (1996), Lee, Rosenzweig and Pitt (1997), and Jalan and Ravallion (2003).

poor, still lacks access to the water network, and there is concern about whether private companies have any incentives to serve extremely poor households.

This chapter examines the effects of a program of expansion of the water network in urban shantytowns undertaken by a privatized water company in Argentina. Relative to the control group, the findings indicate large reductions in the presence, severity, and duration of diarrhea episodes among children in the treated population, as well as reductions in water-related expenses. It turns out that unserved households need to rely on alternative sources of water that require greater monetary expenditures and higher transportation costs than households that have network connections. This chapter also shows that these health and savings effects are important for households that previously had clandestine self-connections to the water network, which were free but of low quality.

Water Privatization: The Case of Aguas Argentinas

Argentina undertook a broad privatization program during the 1990s, which included the provision of water and sanitation services. In 1990, before privatization, public companies provided water service to two-thirds of the country's municipalities, while not-for-profit cooperatives provided service to the remaining one-third. Between 1991 and 1999, about half of the public water companies, serving 28 percent of the country's municipalities and almost 60 percent of the country's population, were transferred to private for-profit control. The largest water company privatization was that of the federal company Obras Sanitarias de la Nación (OSN), which provided service in the Buenos Aires metropolitan area.

In May 1993 Aguas Argentinas, a private consortium led by the French company Lyonnaise des Eaux, won a 35-year concession to provide water service previously provided by OSN.[2] The terms of the concession established service quality and waste-treatment standards.

---

[2] Concessions, rather than the sale of assets to the private firms, are the most common method of privatizing water services worldwide (Noll, Shirley, and Cowan, 2000).

The terms also stipulated construction plans to expand the water network to 100 percent of households and the sewage network to 95 percent of households in metropolitan Buenos Aires by the end of the 35-year period.

Privatization apparently increased efficiency, productivity and investments.[3] Several case studies show large increases in water and sewage production, reductions in spillage, and significant service enhancements. In addition, summer water shortages disappeared, repair delays shortened, and water pressure and cleanliness improved. These service improvements were accompanied by an almost 50 percent reduction in the number of Aguas Argentinas employees. The employment reduction, together with the increase in coverage and production, resulted in large productivity increases. Investments were particularly important in terms of increased access to the network. More than 2 million people in metropolitan Buenos Aires gained access to the water service, and about 1.24 million people gained access to the sewage network. In spite of this significant network expansion, however, more than 15 percent of the population is still not connected to the water network, and more than 40 percent still lacks access to the sewage network. This large unconnected fraction of the population is located in the poorest neighborhoods of metropolitan Buenos Aires.

Although water privatization brought significant progress, if not universal connection, it was not popular in the opinion polls.[4] This was

---

[3] See Artana, Navajas and Urbiztondo (2000); Alcázar, Abdala and Shirley (2002); and Galiani et al. (2005).

[4] The Buenos Aires water privatization did not imply a significant price increase. Indeed, water-use fees in Buenos Aires were initially lowered by 26.9 percent as a result of the privatization bid. However, 13 months after privatization, the regulator authorized a 13.5 percent increase in the usage fee, and a significant increase in connection fees. In response to protests, this high connection fee was later lowered to about one-tenth of the previous level, and a fixed charge was added to the water-use bills for all clients as a cross-subsidy. Some customers also suffered fee increases through property reclassifications (for most customers, fees are based on property characteristics, not on metered use). Indeed, the Buenos Aires water concession has been criticized for its frequent tariff renegotiations. For a general discussion on the evolution of tariffs under privatization in Latin America, see McKenzie and Mookherjee (2003); for the Argentine case, see Alcázar, Abdala and Shirley (2002); Gerchunoff, Greco and Bondorevsky (2003); and Clarke, Kosec and Wallsten (2003).

neither a particular characteristic of the water privatization nor particular to Argentina. Opinion polls and press articles report widespread discontent with privatization in general in Latin America (e.g., IDB, 2002; McKenzie and Mookherjee, 2003), and the Argentine macroeconomic crisis of 2001–02 increased political tensions. The government did not allow the tariff increases stipulated in the contracts, and the private companies interrupted investments. The conflict between Aguas Argentinas and the government escalated until the water concession was cancelled in March 2006, and water provision was transferred to the newly created public company AySA.[5] However, the particular program of water expansion in shantytowns analyzed in this chapter took place entirely before this renationalization.

Expansion of Water in Shantytown Neighborhoods

After 10 years of concession, several shantytown neighborhoods were still not connected to the water and sewage networks. A main regulatory problem for the expansion of the water network in shantytowns is that these areas are not formally urbanized and parcelized. This means that Aguas Argentinas was not licensed to provide service to them. In order to address the specific problem of water development in shantytowns, Aguas Argentinas created a department in 1999 called Development of the Community (*Desarrollo de la Comunidad*) with the objective of designing a methodology to provide water and sewage services to shantytowns. Three years later, at the beginning of 2002, this department launched a specific expansion program, the *Modelo Participativo de Gestión* (MPG, or Participative Expansion Program), in collaboration with neighborhood communities, municipal governments, and the regulatory agency Ente Tripartito de Obras y Servicios Sanitarios (ETOSS).

---

[5] Although only a few of the companies privatized during the 1990s have returned to public hands, Aguas Argentinas was not the only case. Other examples include the mail service, the water company of the Province of Buenos Aires (which returned to provincial administration after Enron's bankruptcy), the public purchase of a fraction of the shares of the national airline company, and the creation of a new public energy company.

The program operated as follows. First, a poor neighborhood community had to ask Aguas Argentinas for provision of the service. Second, the firm evaluated the technical feasibility of the extension of water and/or sewage services to that area. The program did not include the extension of the primary water network, but only secondary connections. Thus, a technical condition was that the neighborhoods had to be less than two blocks away from a covered area. Third, if the provision of the service was technically feasible, Aguas Argentinas asked the neighborhood's municipality for approval to initiate the program in the area. If the municipality approved the project, an agreement among the municipality, Aguas Argentinas, the neighborhood, and ETOSS was signed. This agreement stipulated the role and responsibilities of each party in the program. A salient characteristic of this contract was that beneficiary households agreed to provide the labor for the execution of the construction works in the neighborhood. In return for this work, Aguas Argentinas did not charge water connection fees to any household in the neighborhood. Once connected, the households had to pay a bimonthly service fee of about five pesos (around US$ 1.7) per household.[6] Households also agreed to eliminate all alternative installations of water, including any clandestine connections that might exist.

A clandestine water connection is defined in this chapter as a water connection inside the house that was not provided by the water company. In the sample examined here, about half of the households had a clandestine connection to the water network. It is worth emphasizing that this high percentage is not representative across the Buenos Aires metropolitan area, but was the result of the particular fact that the program was restricted to shantytowns less than two blocks away from connected neighborhoods. Because clandestine connections cannot be carried out over long distances, they are particularly developed in close-by shantytowns, where it is also very difficult for water company personnel to get safe access to the connections to shut them down.

---

[6] The standard connection fee at that time was 138.4 pesos, and the bimonthly service fee was 11 pesos. According to Aguas Argentinas, their labor costs for these connections would have amounted to 132.5 pesos.

The MPG was an institutional arrangement in which public institutions (the water regulatory agency and the municipal government), a private firm, and the community came together to provide piped water and/or sewage services to poor neighborhoods. The program was institutionalized through a contract among Aguas Argentinas, the neighborhood, and the neighborhood's municipality. ETOSS supervised and authorized the whole process.

It is important to consider this arrangement in the context of Argentina's macroeconomic situation. During the 1990s, the country began massive market reforms. However, the performance of labor markets during that period was very disappointing. Poverty and unemployment increased and income distribution worsened (see Altimir and Beccaria, 1998). Social conditions sharply deteriorated during the 2001–02 crisis, when Argentine GDP plummeted, unemployment exploded, and significant inflation affected the purchasing power of Argentine households. At the end of 2003, when the research for this chapter began, around 6 million people—of a total population of 12.6 million in Greater Buenos Aires—were living on less than 3 dollars per day.[7] Of those living in poverty, 44 percent were considered indigent, meaning they lived on a dollar per day or less. Although the number of people below the poverty line began to decline, it still remained very high at around 31 percent by the end of 2005, in spite of strong GDP growth—around 9 percent—in 2003, 2004 and 2005. These figures highlight the enormous magnitude of the poverty problem in Argentina.

According to Aguas Argentinas,[8] by 2002 there were 593 shantytown neighborhoods with an estimated population of 2.5 million people within the concession area. Of this total, 445 neighborhoods with an estimated population of 1.1 million people were within the limits covered by the water network but were without service because of the lack of urbanization and parcelization. The Aguas Argentinas MPG program was a partial attempt to overcome this problem by developing the provision of water and/or sewage service to shantytowns located in the concession

---

[7] See INDEC (2003, 2004).
[8] See Botton, Braïlowsky and Matthieussent (2003).

area. The program would also improve the public image of the company by showing its concern for the poor. The conditions for participating in the program were:

- The shantytown community had to request the service. The neighbors had to be organized and they had to elect representatives to form a community assembly to interact with the other actors in the program. The community also provided labor for construction.
- The neighborhood municipality contractually agreed to undertake certain activities: assigning a supervisor of works; ordering the opening of streets (if required); assigning, if necessary, people with assistential work plans to carry out construction work in front of churches, health centers, schools, and other community buildings in the neighborhood; and distributing work tools (gloves, spades, etc.).
- Aguas Argentinas had to guarantee the technical feasibility of the project. It agreed to deliver the necessary materials (pipes, keys, etc.) and to assure the technical formation of the labor force through several training workshops on work techniques and aspects of labor security. The company was also responsible for communicating the commercial aspects of the service to the community's assembly. It additionally organized social workshops for families in the neighborhoods in order to develop a good relationship between the new clients and the firm, and to provide information about potential health gains from water infrastructure improvements and benefits from responsible use of the service.

After the program's connection works were finished, the households in the neighborhood were incorporated as clients of Aguas Argentinas, and each household had to pay a bimonthly service fee of about five pesos.

In 2003, 33 MPG projects were approved by the water company and the respective municipal governments. As of October 2003, agreements had already been signed for 17 of these projects. Of these, six MPG projects were already finalized, nine were ready to begin, and two were signed, though without a scheduled starting date. Agreements for the remaining 16 projects had not yet been signed (though all of the 33 MPG projects were eventually executed by the program). Table 2.1 shows these projects divided by location. The different progress of these projects may have been due to the different times at which the neighborhoods presented their requests (which, in turn, may reflect different organizational structures), to delays in obtaining approval and construction tools from the municipality, or to delays in obtaining technical approval, construction materials, and workshop developments from the company. The different timing may also be related to geographic neighborhood characteristics, such as the distance to the water network or the need to open a new street.

Sampling Design

In order to study the impact of these MPG projects on the health and expenditures of poor households, a before-and-after study was undertaken comparing the performance in treated neighborhoods relative to a control group. Since the MPG programs began with a community's request for service connection, we selected as the control group neighborhoods that had asked Aguas Argentinas for water service, but that for administrative reasons were not included in the program. The before-and-after strategy permits the use of household fixed effects to address the potential presence of time-invariant differences between treatment and control groups.

For the treatment group, of the 33 projects in Table 2.1, this chapter considers only the 27 neighborhoods where the MPG programs were not finalized by the beginning of the first survey (mid-February 2004), with a further focus on the 25 neighborhoods that had requested connection only to the water network, not the sewage network.[9] Six neighborhoods from this group were sampled, stratifying the projects by region and choosing neighborhoods that belonged to different municipalities throughout Greater Buenos Aires; these neighborhoods are listed in Table 2.2.

---

[9] Sewage construction takes much longer, impeding comparability. Two neighborhoods were excluded for this reason.

Table 2.1	MPG Projects (as of October 2003)				

| | | | Number of | Status of Work | |
| | Number of | | Agreements | Ready to | |
Location	Projects	Population	Signed	Begin	Finalized
South Region GBA	10	18,320	5	4	1
North Region GBA	15	9,350	12	5	5
West Region GBA	4	3,445	0	0	0
Buenos Aires City	4	18,832	0	0	0
Total	33	48,947	17	9	6

Source: "Hacia un acceso universal a los servicios. El modelo participativo de gestión."
Note: "Ready to Begin" means that the contract agreement among the municipality, the neighborhood, Aguas Argentinas and ETOSS was already signed and the training workshops and social activities were scheduled but the works had not yet begun.

Table 2.2	Neighborhoods in the Treatment Group			

| | | Number of Households | | Date of Finalization |
Region	Neighborhood	Population	Sample	of the MPG Program
North	San Jose	107	41	July 2004
North	Cina Cina	400	70	September 2004
West	San Miguel	209	50	April 2004
West	Hipólito Yrigoyen	198	40	September 2004
South	La Tablada	360	70	April 2004
South	10 de Enero	271	95	May 2004
Total		1,545	366	

In addition to these six neighborhoods participating in the MPG program, three neighborhoods were selected to form a control group. These three control neighborhoods were chosen from a group of seven neighborhoods that had asked Aguas Argentinas for water service through the MPG project but were not included in the program for administrative reasons. Table 2.3 presents information on these neighborhoods.

Once all neighborhoods were defined, a systematic household survey of each neighborhood was conducted. The pollsters had a sketch of each neighborhood and a previously defined pathway, guaranteeing that the entire neighborhood was covered. The survey included questions regarding the incidence of water-related illnesses among young children

Table 2.3	Neighborhoods in the Control Group		
		Number of Households	
Region	**Neighborhood**	**Population**	**Sample**
North	Villa Lanzone	435	94
North	Villa Hidalgo	554	95
South	La Rivera	120	80
Total		1,109	269

(less than 6 years old),[10] hygiene habits, trips to medical centers, medical consultations related to water-related illnesses, and household socioeconomic characteristics.

The baseline survey was conducted in the last two weeks of February and the first week of March 2004, and the follow-up survey was conducted in the same weeks of 2005. The last column in Table 2.2 indicates the date of finalization of the program in each neighborhood.[11]

In October 2004, the regular Aguas Argentinas water network expansion reached the "La Rivera" neighborhood and, as a consequence, some of the sampled households in that control neighborhood gained access to water service by November of that year. Since the first survey had already been conducted in that neighborhood, the treatment and control observations were redefined accordingly, transferring the connected households from the control group to the treatment group.[12]

In order to compare the pre-treatment characteristics of the treated and control groups, Table 2.4 shows mean equality tests for

---

[10] The incidence was taken for the month of reference (i.e., one month before the survey was conducted).

[11] Before the survey, it was essential to contact some key people in each neighborhood who could guarantee the safety of the pollsters. These key people were community leaders, schoolteachers, or priests in the neighborhood. In order to have a low nonresponse rate, we agreed with these leaders to donate educational material to each neighborhood school.

[12] As shown in Appendix B of the working paper on which this chapter is based (Galiani, González-Rozada and Schargrodsky, 2007), the results are robust to excluding from the analysis all the observations from the "La Rivera" neighborhood, and they are also robust to excluding only those households that received water in that neighborhood.

	Table 2.4	Mean Equality Tests		

Variable	Treatment Group	Control Group	Diff
Gender of head of household (HH)	0.778	0.804	0.026
	(0.02)	(0.029)	(0.035)
Education of HH (years)	7.347	7.284	−0.063
	(0.125)	(0.21)	(0.244)
Marital status of HH (married=1)	0.721	0.778	0.057
	(0.021)	(0.03)	(0.037)
Nationality of HH (foreigner=1)	0.134	0.093	−0.041
	(0.016)	(0.021)	(0.026)
Employment of HH (employed=1)	0.89	0.88	−0.01
	(0.017)	(0.027)	(0.031)
Unemployment of HH (unemployed=1)	0.11	0.12	0.01
	(0.017)	(0.027)	(0.031)
Age of HH	41.347	45.314	3.967**
	(0.679)	(0.895)	(1.123)
House has a bad roof	0.057	0.066	0.009
	(0.012)	(0.018)	(0.022)
House has electricity	0.973	0.985	0.012
	(0.008)	(0.009)	(0.012)
Water used for cooking is from clandestine connection	0.519	0.554	0.035
	(0.024)	(0.036)	(0.043)
Water used for personal hygiene is from clandestine connection	0.519	0.549	0.03
	(0.024)	(0.036)	(0.043)
Water used for drinking is from clandestine connection	0.482	0.477	−0.005
	(0.024)	(0.036)	(0.043)
Per capita family income	140.02	120.07	19.948**
	(8.35)	(4.86)	(9.66)

** Significant at the 5 percent level.

several variables using the survey conducted in February and March 2004. Both groups share similar characteristics in terms of houses and heads of households, which made treatment and control groups comparable. There is a 20-peso difference between the per capita family income[13] of control and treatments groups, statistically significant at the 5 percent level.

---

[13] For those families in both groups that did not report their income during the survey, family income was imputed using the hotdeck methodology within neighborhoods, using age, formal education, and gender of the head of household as stratification variables.

Estimation Strategy

The main objective is to identify the impact of the MPG water network expansion on children's health status and on the water-related expenses of these households. Of primary interest is the average effect of access to the water network on water-related illnesses in young children less than six years old. In particular, these water-related illnesses are measured through the presence of diarrhea episodes in young children during the month before the surveys were conducted, the duration of those episodes, and whether those episodes included blood or parasites (as a way to assess their severity).

Given the *ex ante* poor quality of the water provision in the neighborhoods, most households bought bottled water to drink—regardless of whether they had clandestine connections to water services.[14] Therefore of additional interest are changes in water-related expenditures as a result of the MPG program. It should be noted that this effect is very important, since the potential increase in fees under privatization for those people already connected to the service provision has been frequently discussed. However, the savings for those who gained access and had been purchasing water from more expensive sources has been mainly ignored in public debates.

Assessing the impacts of access to the water network can be done by comparing the health or economic output of interest in those neighborhoods where the program has been applied with the same outcome in counterfactual neighborhoods—in other words, in the same treatment neighborhoods at the same point in time, but without the application of the program. Since this counterfactual is not observed, it must be estimated. Although it would be ideal to randomly assign access to the water system across neighborhoods, in the absence of this randomized trial the impacts of interest are estimated using the nonexperimental method of difference in differences.

---

[14] Bottled water comes in 5, 10, or 20 liters, and the retail price of the 10-liter bottle is around 3.60 pesos (US$1.2 at the time of the study). In the baseline, water-related expenditures represented 5.3 percent of average household income.

The difference-in-differences model can be specified as a two-way fixed effect linear regression model:

$$y_{it} = \alpha\, I_{it} + \beta x_{it} + \lambda_t + \mu_i + \varepsilon_{it}$$

where y_{it} is the outcome of interest—such as the presence of diarrhea episodes, duration and severity of these episodes, water-related expenses, etc.—for household (child) i in year t; I_{it} is an indicator variable that takes on a value of 1 if household (child) i received water network treatment in year t; x_{it} is a vector of control variables; λ_t is a time effect; μ_i is a household fixed effect; and ε_{it} is the error term.

Results

First considered is the effect on child health of gaining access to piped water through the water network. Table 2. 5 shows very large reductions in the presence of diarrhea episodes (measured with a dummy variable),

Table 2.5	Health Effects of the Expansion of the Water Network					
Dependent Variable	**Presence of Diarrhea Episodes**		**Duration of Diarrhea Episodes**		**Severity of Diarrhea Episodes**	
	(1)	(2)	(3)	(4)	(5)	(6)
I_{it}	−0.1047**	−0.1274*	−0.9359	−1.1241	−0.0644**	−0.0814*
	(0.0525)	(0.0737)	(0.4707)	(0.6581)	(0.0320)	(0.0488)
	{0.0464}	{0.0691}	{0.5383}	{0.7938}	{0.0339}	{0.0524}
% change	−74.26	−90.35	−84.13	−101.04	−87.38	−110.45
Per capita income		−0.0001		−0.0003		0.0003
		(0.0004)		(0.0036)		(0.0004)
		{0.0004}		{0.0036}		{0.0005}
Household fixed effects	Yes	Yes	Yes	Yes	Yes	Yes
Period fixed effects	Yes	Yes	Yes	Yes	Yes	Yes
Mean dependent variable	0.1410	0.1410	1.1125	1.1125	0.0737	0.0737
Observations	819	649	819	649	819	649
R-squared	0.53	0.56	0.50	0.52	0.46	0.47

Notes: The regressions are run at the child level for children less than six years old. Robust standard errors are in parentheses. Robust standard errors clustered at the household-period level to address potential correlation among children of the same household are in curly brackets. The percentage change is computed using the sample average of the dependent variable in the first survey as the baseline level. Statistical significance calculated using the robust standard errors in parentheses: * significant at 10 percent, ** significant at 5 percent, *** significant at 1 percent.

the duration of those episodes (measured in days), and their severity (whether the episodes included blood and/or parasites). The effects are always significant at conventional levels. Controlling for changes in household income, diarrhea episodes decrease by as much as 90 percent of the baseline incidence. Besides reducing the number of diarrhea episodes, access to piped water shows reductions in the duration and severity of those episodes. This evidence complements the findings of Galiani, Gertler and Schargrodsky (2005) on child mortality. In this case, the provision of potable water has a positive effect on reducing morbidity.

An important issue is that approximately half of the treated group was connected clandestinely to the water network and was receiving water, but it was of poor quality. This is shown in Table 2.6, which presents reported improvements in water after the treated areas were connected to the water system.

Tables 2.7 and 2.8 present the effect of access to the water network on the distance walked by household members to bring water to the house (Table 2.7) and on water-related expenditures by families (Table 2.8). Table 2.7 uses as dependent variable the distance (measured in 100-meter blocks) traveled by household members to the nearest hand-pumped well in order to bring water to the house. The program shows a statistically significant effect in reducing the distance traveled by household members to bring water to the house.

Finally, the program also contributed to reducing the money families allocated to water-related expenditures. Table 2.8 shows that, on average,

Table 2.6 Changes in the Perception of Water Quality before and after Treatment

| | Treatment Group (in percentages) | | | | | | | |
| | Color | | Taste | | Odor | | Pressure | |
Quality	Before	After	Before	After	Before	After	Before	After
Very good	1.06	16.25	0.71	13.43	0.71	12.37	0.00	19.08
Good	52.65	74.91	53.71	72.08	55.48	73.14	36.75	59.36
Regular	30.04	7.07	28.27	11.31	26.86	11.66	15.55	7.77
Bad	12.37	1.41	13.07	2.47	13.43	2.47	32.86	5.30
Very bad	3.18	0.35	2.47	0.00	2.47	0.00	12.72	8.13
No answer	0.71	0.00	1.77	0.71	1.06	0.35	2.12	0.35

Table 2.7	Effect on the Distance Walked to Procure Water

Dependent Variable: Distance to Well (in Blocks)	(1)	(2)
I_{it}	−0.8303***	−0.8682***
	(0.1492)	(0.1936)
% change	−77.48	−81.02
Per capita income		0.0005
		(0.0007)
Household fixed effects	Yes	Yes
Period fixed effects	Yes	Yes
Mean dependent variable	1.0716	1.0716
Observations	934	724
R-squared	0.97	0.98

Notes: The regressions are run at the household level. Robust standard errors are in parentheses. The percentage change is computed using the sample average of the dependent variable in the first survey as the baseline level. * significant at 10 percent, ** significant at 5 percent, *** significant at 1 percent.

Table 2.8	Effects on Water-Related Expenditures

Dependent Variable	Water-Related Expenditures (in Pesos)		Water-Related Expenditures Including Payment for Water Service after the Program	
	(1)	(2)	(3)	(4)
I_{it}	−21.8678***	−19.7232**	−19.2469***	−17.1078**
	(6.3073)	(8.0001)	(6.2267)	(7.9029)
% change	−94.97	−85.65	−83.59	−74.30
Per capita income		−0.0001		0.0000
		(0.0175)		(0.0174)
Household fixed effects	Yes	Yes	Yes	Yes
Period fixed effects	Yes	Yes	Yes	Yes
Mean dependent variable	23.0266	23.0266	23.0266	23.0266
Observations	483	369	483	369
R-squared	0.85	0.88	0.85	0.88

Notes: The regressions are run at the household level. Robust standard errors are in parentheses. The percentage change is computed using the sample average of the dependent variable in the first survey as the baseline level. * significant at 10 percent, ** significant at 5 percent, *** significant at 1 percent.

families in the treatment group spent about 20 to 22 pesos per month (around US$7 to $7.5) less than families in the control group that bought water. This evidence implies that the provision of potable water through the MPG program reduced water-related expenditures between 85 and 95

Table 2.9   **Actual Spending of the Money Saved from Water-Related Expenditures in the Treatment Group**

Expenditures on:	Percentages
Food/beverages	67.61
Personal items	2.82
Items for children	9.86
Savings/paying public utilities	1.41
General expenses	14.08
No answer	4.23

percent with respect to the sample average of water-related expenditures at the baseline level. These savings represent around 4.5 percent of average total family income in the baseline survey for the treatment group.

Finally, Table 2.9 presents evidence on the allocation of this extra money by treated households. Families connected to the water network allocated most of the water savings to food consumption.

Another important question is whether the expansion of the water network had positive or negative impacts on households in the treatment group that already had a clandestine self-connection to the water network. On the one hand, the expansion program meant that these families had to pay for something they were already receiving for free. On the other hand, these clandestine self-connections are of inadequate quality in terms of water pollution and low pressure. Table 2.10 differentially analyzes the effect of the MPG expansion for previously unconnected and clandestinely connected households. The results broadly show that the health and savings effects are about half for the clandestinely connected than they are for the previously unconnected, but in both cases the effects are large.

Conclusions

This chapter studies the effects of a specific program of water expansion in urban shantytowns undertaken by a privatized company in Argentina, finding large reductions in the incidence, duration and severity of diarrhea episodes among children in the treated population. The results further include a significant reduction in water-related expenses, since piped

| Table 2.10 | Differential Effects for Unconnected Households and Households with Previous Clandestine Connections |

Dependent Variable	Presence of Diarrhea Episodes	Duration of Diarrhea Episodes	Severity of Diarrhea Episodes	Distance to Well (in Blocks)	Water-Related Expenditures
	(1)	(2)	(3)	(4)	(5)
Water for	−0.1358**	−0.6544	−0.0802**	−1.1154***	−29.9020*
unconnected	(0.0607)	(0.4757)	(0.0398)	(0.2146)	(16.9421)
households	{0.0557}	{0.4659}	{0.0452}		
% change	−96.31	−58.82	−108.82	−104.09	−129.86
Water for households	−0.0686	−1.2626*	−0.0461	−0.6079***	−19.6430***
with clandestine	(0.0586)	(0.7392)	(0.0339)	(0.2059)	(5.9026)
connection	{0.0505}	{0.9584}	{0.0336}		
% change	−48.65	−113.49	−62.55	−56.73	−85.31
Household fixed effects	Yes	Yes	Yes	Yes	Yes
Period fixed effects	Yes	Yes	Yes	Yes	Yes
Mean dependent variable	0.1410	1.1125	0.0737	1.0716	23.0266
Observations	819	819	819	934	483
R-squared	0.529	0.505	0.458	0.972	0.851

Notes: The regressions in columns (1) to (3) are run at the child level, whereas columns (4) and (5) are run at the household level. Robust standard errors are in parentheses. Robust standard errors clustered at the household-period level to address potential correlation among children of the same household are in curly brackets. The percentage change is computed using the sample average of the dependent variable in the first survey as the baseline level. These results do not change if per capita income is included as a regressor. Statistical significance calculated using the robust standard errors in parentheses: * significant at 10 percent, ** significant at 5 percent, *** significant at 1 percent.

water is cheaper than bottled water. These health and savings effects are important for households that previously had a clandestine self-connection to the water network, which was free but of low quality.

The results highlight two main findings that are important as policy lessons. First, there are significant savings from gaining access to the water network for those households that previously had to rely on alternative water sources (such as bottled water, clandestine connections or water transported from distant wells). These savings are overlooked in traditional analyses of the privatization of water services. Second, the apparent monetary loss for those households when replacing their free clandestine self-connections with a formal water network connection may be overcome by important health improvements.

CHAPTER THREE

The Impact of Electricity Sector Privatization on Public Health

Martín González-Eiras and Martín A. Rossi

D uring the 1990s, Argentina undertook structural reforms including a privatization program that transferred most of its national, provincial, and municipal state-owned enterprises to private hands. Among these reforms was the privatization of most electricity companies. Provincial governments are responsible for delivering electricity services and not all of them decided to privatize these services; those that did so undertook the process at different times during the decade. More precisely, between 1992 and 1998, 17 electricity companies covering 70 percent of the population were privatized.

There are several studies analyzing the impact of Argentina's electricity privatization program on some general measures of welfare. Chisari, Estache, and Romero (1999) use a calibrated general equilibrium model to assess the welfare gains of utilities privatization for private consumers. They find positive overall welfare gains whose size, and their effect on the distribution of income, strongly depend on the quality of regulation. Benitez, Chisari, and Estache (2003) modify the model of Chisari, Estache, and Romero (1999) to study the impact of utilities privatization on the public sector. They show that better regulation increases consumer welfare but at the expense of public revenues. From a social point of view, consumer gains are higher than the loss in revenue. Delfino and Cesarin (2003) use data from the Greater Buenos Aires area to measure the consumer surplus from the privatization of several public services. They measure the impact both for initial consumers and for newcomers, and find that for some services—electricity among them—welfare changes are positive.

There is also a study by Galiani, Gertler, and Schargrodsky (2005) on the impact of water privatization on child mortality in Argentina, which shows that privatization reduced child mortality, especially in the poorest municipalities that privatized.

The hypothesis presented in this chapter is that service expansions and quality improvements associated with the privatization of electric companies have had a positive effect on health outcomes, particularly among the poor. Privatization increased access to electricity and thus allowed a number of households, whose only constraint was the lack of electricity in their homes, to have a refrigerator; refrigerator use in turn improves nutrition intake. Privatization also reduces the frequency and duration of interruptions per customer, which may reduce the likelihood of food poisoning. Thus, variation in ownership of provincial electricity companies over time and space is used to identify the causal effect of privatization on some measures of public health related to nutrition and food poisoning.

Argentina's Electricity Reform

Though the electricity industry was wholly state and provincially owned at the beginning of the 1990s, more than 80 percent of the generation sector, all of the transmission sector, and about 70 percent of the distribution sectors in Argentina were transferred to private ownership by 1998.[1]

The process started with the division of SEGBA (a firm owned by the federal government) into three companies in 1992. Two of the new companies, EDENOR and EDESUR, each covered half of the city of Buenos Aires and the area of Greater Buenos Aires, while a third firm, EDELAP, covered the area of Greater La Plata. These companies covered almost 40 percent of the population of the country. At that time, the rest of the electricity distribution in the country was carried out by state public companies and small local cooperatives. San Luis was the first provincial

---

[1] For details of the privatization process, see Pollitt (2004) and Galiani et al. (2005).

Table 3.1	Schedule of the Electricity Privatization Program
Year	**Privatized Firms**
1992	EDENOR, EDESUR, and EDELAP
1993	EDESAL
1995	EDELAR, EDESE, EDET, and EDEFOR
1996	EDESA, ESJSA, EDEERSA, EDERSA, EJESA, and EDECAT
1997	EDEA, EDEN, and EDES
1998	EDEMSA

Source: Secretaría de Energía.

government to grant concessions for its distribution of electricity in 1993, followed by Santiago del Estero, La Rioja, Tucuman, and Formosa in 1995. In 1996, ESEBA, the second-largest company after SEGBA, was divided into three firms: EDEA, EDEN, and EDES. The provinces of San Juan, Jujuy, Entre Ríos, Río Negro, Salta, Catamarca, and Mendoza later replicated the process. Today, around 70 percent of the population is served by private companies (Andrés, Guasch, and Foster, 2004). Table 3.1 and Figure 3.1 show the schedule of the privatization program.

The comprehensive nature of the electricity reform in Argentina reflected the poor performance of the sector prior to privatization and the idea that the private sector could help achieve potential efficiency gains

Figure 3.1	Percentage of Provinces with Privatized Electricity Systems

in terms of expanded access and improved service quality. The question is, have these objectives been achieved?

Access to Electricity Services

To identify the effect of privatization on access to electricity, this chapter exploits the fact that all the privatizations in electricity distribution services occurred from 1992 to 1998 (see Table 3.1), and that provincial-level census data on the proportion of households with access to electricity are available for the years 1991 and 2001.

Using the 1991 and 2001 census data, the difference-in-differences estimate is calculated of the impact of privatization on the proportion of households that had access to electricity. Census data are used, instead of connection data from firm sources, to account for the fact that many households had access to the electricity network via clandestine connections.

The difference-in-differences estimator compares the change in the proportion of households with access to electricity in provinces that privatized to the change in the proportion of households with access to electricity in provinces that did not privatize electricity services. Both the city of Buenos Aires and the province of Buenos Aires are excluded from the analysis since about 99 percent of households were already connected to electricity service before privatization. Results from the difference-in-differences estimator, reported in Table 3.2, show a larger increase in the proportion of households connected to electricity services in provinces that privatized than in provinces that did not. The estimated coefficient indicates that the total number of households with access to electricity increased by 2.3 percentage points as a result of privatization.

Indeed, many authors have highlighted the positive impact of privatization on poor households. According to Bouille, Dubrovsky, and Maurer (2002), one of the striking achievements of the early years of Argentine electricity reform was the sharp increase in the number of poor households with electricity supply. This is likely to have had a positive impact on the social welfare of these households, as they often previously lacked electricity for heating, pumping water, and food preservation. As Pollitt (2004) notes, "Many developing countries face problems of improving

Table 3.2	Impact of Privatization on the Proportion of Households with Access to the Electricity Network		
	Proportion of Households Connected in 1991	**Proportion of Households Connected in 2001**	**Difference 2001–1991**
Provinces that did not privatize	0.893	0.943	0.050
Provinces that privatized[a]	0.859	0.933	0.073
			0.023

Note: There is no sample variability when the proportion of households with access to electricity networks for the years 1991 and 2001 is estimated, since these proportions are estimated from Census data.
[a] Buenos Aires is excluded, since about 99 percent of households were already connected to electricity service before privatization.

the access of the poorest while giving financial incentives to companies to supply them. Argentina handled this problem in an economically efficient way. The increase of access to poor consumers was calculated to have yielded large increases in social welfare and be a significant benefit of the restructuring of the sector."

Quality of Service

A focus on service quality requires the consideration of a variety of issues. In Argentina, privately owned electricity distribution firms are responsible for any shortage of supply, regardless of the cause of this shortage. If interruptions reduce the quality of service below the minimum standards set in concession contracts, then distribution firms have to pay penalties. Concession contracts specify minimum standards in technical product (voltage variations), in technical service (duration and frequency of interruption), and in commercial service (customer complaints and the like). Thus, privately owned electricity distribution firms have strong incentives to provide an adequate quality of service.

Table 3.3 presents some statistics regarding two widely used measures for quality of service: mean frequency of interruptions per customer (FC), defined as

$$FC = \frac{\sum_{i=1}^{n} Ca_i}{Cs}$$

Table 3.3	**Quality of Service**			
	Mean Frequency of Interruption per Customer (FC)		**Total Time of Interruption per Customer (TC)**	
	Number of Observations	**Average**	**Number of Observations**	**Average**
Public firms	10	40.79	9	20.51
Private firms	44	6.10	44	9.69
Total	54	12.52	53	11.52
Before privatization	8	14.15	8	21.72
After privatization	38	6.00	38	9.74

Source: Authors' calculations. Public firms: EPESF (2001), EDELAP (1991, 1992), EDENOR (1992), EDESUR (1992), SECHEEP (1992, missing information on TC), EDEERSA (1996), and EDEMSA (1995–1997). Privatized firms: EDELAP (1993–2001), EDENOR (1993–2001), EDESUR (1993–2001), EDEERSA (1997–2002), EDEMSA (1998–2002), and ESJ (1997–2002).

(where Ca_i is the number of customers affected by interruption i, Cs is the total number of customers, and n is the total number of interruptions), and total time of interruption per customer (TC), defined as

$$TC = \frac{\sum_{i=1}^{n} Ca_i \times t_i}{Cs}$$

(where t_i is the duration of interruption i). As shown in Table 3.3, the averages of both FC and TC for private firms are lower than the average for public firms, which is in line with the idea that private firms have better quality indicators than public firms.[2]

Similar conclusions are reached when considering the before/after performance (in terms of quality of service) of firms that were privatized. Information on TC and FC is available in at least one year before and after privatization for five firms: EDENOR, EDESUR, EDELAP, EDEMSA, and EDEERSA. As shown in Table 3.3, TC decreased for this group of firms from an average of 21.72 before privatization to 9.74 after privatization. FC similarly decreased from 14.15 before privatization to 6.00 after privatization.

The number of hours of supply lost per year provides additional evidence for the increase in the quality of service after privatization. For

---

[2] Of course, this does not mean to suggest that the relation is causal.

the three distribution utilities in the Greater Buenos Aires area (the only utilities for which these data are available) the number of hours of supply lost per year was 21 in 1988, 16.8 in 1993/94, and 5 in 2000/01.

Finally, when an interruption takes place it can be caused at the generation, transport, or distribution stage. One concern regarding the identification strategy employed here was whether distribution is an important factor in explaining service quality. To explore the validity of this concern, many specialists in the electricity sector were consulted, and they agreed that distribution must be considered as an important factor in electricity interruptions. Indeed, information on minutes of interruptions and number of customers suffering interruptions in EDELAP's concession zone in 1999 and 2000 shows that 96 percent of the minutes per customer interrupted and 87 percent of customers suffering interruptions were caused by problems originating in the distribution stage.

In short, there is evidence that the privatization programs of the electricity sector have had an important impact on increasing both access to service and quality of service.

Food Quality, Nutrition, and Public Health

The quality of food intake and health go hand in hand. Up to one-third of people in developed countries are affected by food-borne diseases every year. This problem is likely to be even more widespread in developing countries, because the poor are more susceptible to ill-health. In addition to food contamination, unbalanced diets lead to worsening health outcomes, especially in the form of growth retardation and poor cognitive development of children. This chapter focuses on these two problems and their relationship to the quality of food intake, nutrition, and food poisoning, and their impact on public health.

Nutrition and Low Birth Weight

A diversity of foods in a balanced diet improves nutritional status and health, thus providing another channel through which food impacts public health. Malnutrition and nutrition-related chronic conditions (ischemic heart disease, high blood pressure, and stroke, among others) are more prevalent among the poor. And although low-income households are usually efficient in feeding themselves with little resources, they spend heavily on energy-dense foods (Nelson, 1999) and do not necessarily obtain a balanced diet.

For younger children, an unbalanced diet results in growth retardation and poor cognitive development. Thus, the availability of a refrigerator should have a positive impact on the development of younger children by allowing them access to a more balanced diet. And better nutrition of the mother before and during pregnancy should reduce the probability of low birth weight.

Low birth weight remains a significant public health problem in developing countries since it not only increases infant mortality rates, but also carries long-term risk in the form of high rates of adult coronary heart disease and diabetes (see Barker, 1998). Recently Almond, Chay and Lee (2005) measured the benefit in the United States of an additional pound of weight at birth for babies weighing 2000–2100 grams to be $10,000 in saved hospital charges for inpatient services.

Studies on mothers' micronutrient consumption have shown that in addition to caloric intake, some micronutrients have a positive effect on birth weight. Mardones-Santander et al. (1988) and Rao et al. (2001) have shown with Chilean and Indian data that mothers' consumption of milk fortified with folic acid and iron had a positive effect on birth weight. More recently, Ramakrishnan (2004) surveyed the literature and found little evidence of positive effect of multivitamin mineral supplements, beyond iron and iron-folate supplementation, on birth weight.

Of course, nutrition is not the only cause of low birth weight. Cross-sectional birth weight variation is directly or indirectly influenced by immutable factors (genetics), socioeconomic factors (education, income), maternal behavior beyond nutrition (e.g., smoking behavior), and other environmental factors (such as infections).

Food-Borne Diseases and Diarrheas

Like water and sanitation, the major health burden arising from food contamination is almost certainly its contribution to diarrhea and dysentery, which figure so highly in the morbidity and mortality of children in developing countries. There is also growing evidence of the serious long-term health effects of food-borne hazards, including kidney failure, reactive arthritis, and disorders of the brain and nervous system (World Health Organization, 2001). Food-borne diseases thus take a major toll on health, which has led the World Health Organization and its member states to recognize food safety as an essential public health function.

Epidemiological studies give little indication of the relative importance of food contamination. An attempt to indirectly estimate the share of diarrhea resulting from food contamination placed this figure between 15 percent and 70 percent (Esrey and Feachem, 1989). Since food safety must be addressed along the entire food chain, home food handling and storage practices are critical for preventing contamination.

Food contamination takes place when germs infect noncontaminated foods, or when already present colonies of germs develop under favorable conditions. The most common food-borne diseases are salmonella, shigella, and staphylococcus. All of them, and most other food-borne diseases, are less likely to occur if the cold chain is preserved (foods kept at more than 5 degrees Celsius for as little as two hours can become contaminated), and if foods are cooked, at temperatures above 70 degrees Celsius. Food contamination symptoms start a few hours after ingestion, and always include fever and diarrhea. Many food-borne illnesses are the result of improper storage, and when electricity fails, refrigerated storage is a casualty, with an increased risk of illness. It is therefore likely that food poisoning cases peak as a result of power cuts.

Impact of Electricity Privatization on Health in Argentina

There are two main potential pathways by which the privatization of the Argentine electricity sector might have positively affected public health. The first is that the privatization fostered the expansion of the network, providing access to service to households that were not previously connected to electricity. The second is the improvement in service quality in terms of fewer shortages of supply.

There are at least two channels through which the expansion of service connections (and the subsequent increase in refrigerator use among low-income households) and the improvement of service quality may have had a positive impact on public health. First, an immediate effect arises from the abatement of food-borne diseases. Furthermore, the richer nutritional contents resulting from a more varied diet should improve mothers' micronutrient consumption, and reduce the negative impact of growth retardation and poor cognitive development among children. This second channel is related to service expansion, while the first is related both to service expansion and quality improvements, given that fewer power cuts imply fewer instances of breaks in the cold storage chain.

Of course, low-income families may not be able to afford a refrigerator (or other hygiene aids such as insect- and rodent-proof storage containers). But in the absence of electricity they have no choice but to live without one since there is no good substitute for an electric refrigerator.[3]

These concerns thus make it necessary to consider the impact of electricity distribution privatization and its subsequent network expansion and service quality improvements on health indicators related to food contamination and dietary diversity. The effects observed, if any, would come from low-income families having access to a refrigerator and a lower frequency of food-borne diseases due to breaks in the cold storage chain. The health outcome used to test the "access" channel is the frequency of low-birth-weight births. The "quality" channel is tested by looking at how privatization affected hospitalizations and death rates for causes related to diarrheas and food poisoning. As mentioned above, it is likely that service expansion has also affected this outcome, so it will be impossible to separate the two channels.

Another channel by which a blackout may impact health is through the provision of water. A blackout compromises the water supply in at least two ways: first, by decreasing the pressure in water pumps, allowing

---

[3] Kerosene- or gas-powered refrigerators are possible substitutes, but they are more expensive than electric ones. They were mainly used by medium- to high-income rural families. Survey data for the years 2004 and 2005 in Argentina shows that about 17 percent of households without access to electricity have a refrigerator. This figure is 87 percent for households with access to electricity.

bacteria to build up in municipal water systems; and second, by effectively shutting down sewage treatment facilities.

These factors were evident during an important blackout that cut power to tens of millions of people in the northeastern United States and eastern Canada in August 2003. During the blackout, New York City's Department of Health and Mental Hygiene detected a higher than usual number of visits for diarrheal illnesses at emergency departments in the city. Health Commissioner Thomas Frieden said, "While we do not know the specific cause of this spike in diarrheal illnesses, it is possible that it was caused by spoiled food eaten at home or elsewhere." Also, Detroit-area food poisoning claims skyrocketed during the power outage. This anecdotal evidence highlights the potential health impact of inadequate electricity service.

Data

The data cover the period 1990 to 2000. Given that the first privatization in the electricity sector in Argentina was in 1992 and the last one was in 1998, this 11-year period includes two years before the first privatization and two years after the last privatization. Extending the dataset beyond 2000 might not be appropriate given the macroeconomic crisis faced by the country. Argentina experienced a negative shock when Brazil devalued its currency in 1999, and conditions deteriorated significantly after June 2001, with output falling at a rate of more than 10 percent over the following year and a half. At the beginning of January 2002, the Argentine government defaulted on its debt and sanctioned the Emergency Law, by which tariffs on utilities were "pesified," or converted to pesos and contracts renegotiated. Concretely, Article 8 of the Emergency Law required that tariffs previously stated in U.S. dollars be converted to Argentine pesos at a rate of one-to-one (the previous exchange rate) and that they could no longer be indexed to foreign inflation. To have an idea of what this means in terms of the possibility of investments by private firms, in the six-month period following the pesification law, the Argentine peso went from being worth one U.S. dollar to being worth about 25 U.S. cents. The tariff freezing meant that utility rates, measured in dollars, fell by up to 66 percent.

The main output measures considered are very low birth weight and low birth weight, defined as the frequency of birth with weights below 1,500 grams and 2,500 grams, respectively. The latter measure is a standard in the health literature to gauge the importance of nutritional problems among newborns. It should be noted that there are some years/ provinces for which the number of unrecorded births is high. Another output measure used in the empirical section, Diarrhea and Food Poisoning, is the rate of mortality caused by intestinal infections for children under 5 years of age.

The dataset also includes a privatization dummy variable that takes the value of 1 if electricity services are provided by a private company and 0 otherwise, and a set of province characteristics.[4] The definitions and sources of all variables used in the empirical section are presented in Table 3.4.

Results

The objective is to identify the impact of privatization on measures of public health related to food contamination and nutritional deficiencies in those provinces where the electricity sector has been privatized.

The first set of estimates is obtained using the difference-in-differences estimator, which compares the change in health outcomes for those provinces that privatized their electricity services to the change in health outcomes for those provinces that did not privatize their electricity services.

Formally, the difference-in-differences model can be specified as

$$Y_{it} = \beta D_{it} + \lambda X_{it} + \alpha_i + \mu_t + \varepsilon_{it} \tag{1}$$

where Y_{it} is the output of interest (low birth weight) in a given province in period t, X_{it} is the vector of the subset of control variables in the vector X that vary both across units and time, D_{it} is a dummy variable that takes

---

[4] In some provinces there are cooperative firms providing electricity. In all provinces that privatized their electricity services, the privatized firm serves more than 50 percent of all customers.

Table 3.4	Data Sources and Definitions	
Variable	Definition	Source
Low birth weight	Proportion of infants born weighing less than 2500 or 1500 grams	Ministerio de Salud
Diarrhea and food poisoning	Child mortality rate caused by intestinal infections	Ministerio de Salud
Private	Dummy variable that equals 1 if the largest fraction of the population has electricity services provided by a private company, and 0 otherwise	Secretaría de Energía
Unemployment rate	Unemployment rate (May and October average) for households in the surveyed cities of the province. There is no record for the province of Rio Negro	Permanente Household Survey, INDEC
Real GDP per capita	Per capita gross geographic product in hundreds of constant pesos	Permanente Household Survey, INDEC
Income inequality	Gini index (May and October average) for households in the surveyed cities of the province	Permanente Household Survey, INDEC
Public spending per capita	Current public spending per capita in hundreds of constant pesos (1993)	INDEC
Peronist	Dummy variable that equals 1 if the province is governed by the Peronist party, or if the company providing electricity services depends on the federal government, and 0 otherwise	Ministerio del Interior
Share of water privatization	Proportion of the population in the province with privatized water services	Galiani, Gertler, and Schargrodsky (2005)
Temperature	Number of days with a high temperature above 30°C	CIM, Servicio Meteorológico Nacional

the value of 1 if province i's electricity system was privatized during period t, α_i is a time-invariant province effect, μ_t is a time effect common to all provinces in period t, and ε_{it} is a province time-varying error distributed independently across provinces and time and independently of all α_i and μ_t. The parameter of interest, β, is the difference-in-differences estimate of the average effect of privatization on low birth weight.

The difference-in-differences model assumes that the change in low birth weight in control (not-privatized) areas is an unbiased estimate of the counterfactual. While this assumption cannot be tested directly, it is possible to test whether time trends of low birth weight in provinces that privatized and provinces that did not privatize electricity services were the same in the pre-privatization periods. If time trends are the same in the pre-intervention periods, then it is likely that they would have been the same in the post-intervention period had treated provinces not privatized. As in Galiani, Gertler, and Schargrodsky (2005), a model like that in Equation (1) is estimated to formally test the hypothesis that the pre-intervention time trends for provinces that privatized and did not privatize their electricity services are not different; this model, however, excludes the privatization dummy variable and includes separate year dummies for (eventual) treatments and controls. The data include only observations of the control and the treatment provinces before privatization; that is, data for the years 1990 to 1997 are used for all the control provinces and only the pre-privatization years for those provinces that privatized electricity services (as noted above, the last privatization was in 1998). Since the dummy variables capturing the interaction between the year effects and the eventually privatized electricity systems are not significant at conventional levels of significance, it is not possible to reject the hypothesis that the pre-privatization year dummies are the same for provinces that did privatize and provinces that did not privatize, thus validating the difference-in-differences identification strategy.

Table 3.5 presents difference-in-differences estimates of the privatization of electricity services on the proportion of very low weight births (less than 1,500 grams). The privatization dummy lagged one period is used, since nutrition over the whole pregnancy affects weight at birth. In order to account for the presence of a common random effect at the year-state (public or private) level, standard errors are clustered at the year-privatized/not-privatized level (see Moulton, 1990).

The first column reports the difference-in-differences model without controls, which shows a negative and significant association between privatization and proportion of very low birth weights. The magnitude of the coefficient indicates that privatization is associated with a 0.0021 decrease in the proportion of very low birth weight, which corresponds to a 21 percent reduction of the baseline proportion.

Table 3.5 Difference-in-Differences Estimates of the Impact of Privatization of the Electricity Sector on Low Birth Weight

| | Dependent Variable: Birth Weight | | | | | | | | | |
| | Proportion less than 1,500 g | | | | | | Proportion less than 2,500 g | | | |
	(1)	(2)	(3)	(4)	(5)	(6)	(7)	(8)	(9)	(10)
Private electricity service$_{t-1}$	-.0021	-.0023	-.0024	-.0020	-.0001	-.0027	-.0043	-.0043	-.0036	-.0007
	(.0008)**	(.0009)**	(.0010)**	(.0010)*	(.0004)	(.0019)	(.0017)**	(.0019)**	(.0018)*	(.0007)
	[.0026]	[.0027]	[.0027]	[.0025]	[.0004]	[.0053]	[.0056]	[.0057]	[.0053]	[.0012]
% Δ in proportion of low birth weight	-21.37	-23.44	-24.36	-20.25	-1.30	-3.70	-5.77	-5.78	-4.91	-0.93
Ln(Real GDP per capita)		-.0030	-.0026	-.0029	.0008		.0157	.0157	.0153	.0014
		(.0039)	(.0037)	(.0037)	(.0020)		(.0165)	(.0160)	(.0162)	(.0041)
		[.0054]	[.0052]	[.0046]	[.0014]		[.0136]	[.0135]	[.0139]	[.0042]
Unemployment rate		-.0239	-.0230	-.0213	-.0027		-.0360	-.0358	-.0331	-.0196
		(.0078)***	(.0077)***	(.0087)**	(.0048)		(.0364)	(.0349)	(.0351)	(.0103)*
		[.0194]	[.0189]	[.0183]	[.0063]		[.0423]	[.0416]	[.0407]	[.0111]*
Income inequality		.0826	.0831	.0784	-.0038		.1922	.1923	.1848	.0136
		(.0372)**	(.0380)**	(.0353)**	(.0061)		(.0763)**	(.0764)**	(.0735)**	(.0121)
		[.0658]	[.0659]	[.0624]	[.0068]		[.1171]	[.1177]	[.1120]	[.0220]
Ln(Public spending per capita)		.0059	.0055	.0057	.0001		.0081	.0081	.0083	-.0015
		(.0053)	(.0053)	(.0053)	(.0010)		(.0081)	(.0083)	(.0081)	(.0033)
		[.0050]	[.0049]	[.0050]	[.0014]		[.0083]	[.0084]	[.0088]	[.0033]
Province governed by Peronist party			-.0008	-.0005	-.0001			-.0001	-.0002	-.0007
			(.0005)	(.0006)	(.0004)			(.0021)	(.0023)	(.0012)
			[.0008]	[.0007]	[.0004]			[.0015]	[.0014]	[.0008]
Share of water privatization					.0038	-.0002			-.0002	.0004
				(.0023)	(.0003)				(.0041)	(.0011)
				[.0025]	[.0004]				[.0051]	[.0016]
Observations	242	230	230	230	218	242	230	230	230	218

Notes: All regressions include year and province fixed effects, and exclude Buenos Aires since about 99 percent of households were already connected to the service before privatization. Mean values of low birth weight in 1990 are 0.010 (<1,500 g) and 0.076 (<2,500 g). Standard errors clustered at the year-private level are in parentheses. Standard errors clustered at the province level are in brackets. *Significant at the 10 percent level; **Significant at the 5 percent level; ***Significant at the 1 percent level.

One concern regarding this type of study is that there may be time-varying province characteristics correlated with both the weight at birth and the electricity sector being in private hands. To address this concern, Column 2 controls for a number of observed time-varying characteristics, including GDP per capita, unemployment, income inequality, and public spending per capita. The coefficients on GDP per capita and public spending per capita are not significant at any of the usual confidence levels. As expected, income inequality has a positive and significant sign, suggesting that a worse distribution of province income is associated with a higher proportion of very low birth weights. The coefficient on unemployment rate is negative and significant, which is a puzzling result. The coefficient on the lagged privatization dummy remains negative and significant at the 5 percent level.

Column 3 adds a dummy variable for the political party controlling the local government in order to control for political preferences for health outcomes beyond public spending levels. As reported in Column 3, the added variable is not significant and does not have any impact on the estimated coefficients and significance of the other variables.

An additional concern is that the same provinces that privatized electricity services might have also privatized water services, and that water privatization rather than electricity privatization is responsible for the decrease in the proportion of very low birth weights. To address this concern, the privatization of water services is controlled for by including the proportion of the population in the province with privatized water services as an additional regressor.[5] As reported in Column 4, the coefficient associated with the privatization of water services is not significant. The coefficient of the lagged privatization dummy remains negative and significant at the 10 percent level.

As shown in Columns 6–9, similar results are obtained when low birth weight is defined as the proportion of births under 2,500 grams instead of the proportion of births under 1,500 grams. The magnitude of the coefficients indicates that privatization is associated with a decrease

---

[5] Municipalities, not provinces, are responsible for delivering water services.

in the proportion of low birth weight in the range of 0.0028 to 0.0044, which corresponds to a reduction of the baseline proportion in the range of 3.7 percent to 5.8 percent.

To conclude, two caveats should be considered when interpreting these results. First, as pointed out by Bertrand, Duflo, and Mullainathan (2004), difference-in-differences estimates may suffer from a potential problem of serial correlation of the error term. To avoid potential biases in the estimation of the standard errors arising from serial correlation, an arbitrary covariance structure within provinces over time is allowed for by computing standard errors clustered at the province level. When standard errors corresponding to the estimates reported in Table 3.5 are computed in this way, the coefficients on privatization are not significant at conventional levels of confidence. This may be related to the small number of cross-section observations (i.e., provinces) available.

Second, in some years and for some provinces there is a large proportion of infants not weighed at birth. As shown in Columns 5 and 10, when those observations where the proportion of infants weighed at birth is less than 70 percent (12 observations) are excluded from the sample, the coefficients on privatization become not significant at conventional levels of confidence. More worrisome is the fact that the magnitude of coefficients changes dramatically, suggesting that these observations may be driving previous results.

Overall, difference-in-differences estimates provide weak evidence that privatization is negatively associated with low birth weight.

Food Poisoning

Table 3.6 presents the results of the food-poisoning pathway by which privatization might have a positive impact on health. The dependent variable is the child mortality rate caused by diarrhea and food poisoning. It is measured as the ratio of the number of deaths caused by diarrhea and food poisoning in children under 5 years of age to the total number of children under 5 alive at the beginning of the year. The figures focus on young children because they are particularly vulnerable to diseases related to food poisoning as a result of weak body defenses.

As reported in Column 1 of Table 3.6, in the difference-in-differences model without controls there is a negative though not significant association between the two variables.

Table 3.6	**Difference-in-Differences Estimates of the Impact of Privatization of the Electricity Sector on Child Mortality Rates Caused by Diarrhea and Food Poisoning**

	Dependent Variable: Child Mortality Rates Caused by Diarrhea and Food Poisoning				
	(1)	(2)	(3)	(4)	(5)
Private electricity service$_{t-1}$	−.000015	−.000078	−.000091	−.000088	−.000083
	(.000015)	(.000025)***	(.000033)**	(.000032)**	(.000036)***
	[.000027]	[.00009]	[.00008]	[.000080]	[.000068]
Temperature		-5.85e-07	-6.00e-07	-5.06e-07	-4.87e-07
		(4.01e-07)	(4.32e-07)	(3.97e-07)	(3.90e-07)
		[5.18e-07]	[5.57e-07]	[5.39e-07]	[5.13e-07]
Temperature*Private electricity service$_{t-1}$		6.41e-07	8.24e-07	7.50e-07	7.18e-07
		(1.95e-07)***	(2.99e-07)**	(2.73e-07)**	(2.37e-07)***
		[7.17e-07]	[7.53e-07]	[6.93e-07]	[6.27e-07]
Ln(Real GDP per capita)			.00020	.00022	.00022
			(.00006)***	(.00007)***	(.00007)***
			[.00010]*	[.00010]**	[.00010]**
Unemployment rate			−.00016	−.00012	−.00011
			(.00018)	(.00019)	(.00019)
			[.00037]	[.00038]	[.00039]
Income inequality			−.00051	−.00049	−.00052
			(.00025)*	(.00027)*	(.00026)*
			[.00055]	[.00053]	[.00053]
Ln(Public spending per capita)			.00018	.00017	.00017
			(.00008)**	(.00008)*	(.00008)*
			[.00018]	[.00018]	[.00018]
Province governed by Peronist party				−.00004	−.00004
				(.00002)*	(.00002)*
				[.00002]*	[.00002]*
Share of water privatization					−.000017
					(.00002)
					[.000041]
Observations	264	264	252	252	252

Notes: All of the regressions include year and province fixed effects. The mean values of child mortality rate caused by diarrhea and food poisoning in 1990 is 0.00025. Standard errors clustered at the year-private level are in parentheses. Standard errors clustered at the province level are in brackets.
*Significant at the 10 percent level; **Significant at the 5 percent level; ***Significant at the 1 percent level.

To further explore the association between privatization and child mortality rate caused by diarrhea and food poisoning, Column 2 includes the number of days with temperatures above 30 degrees Celsius and its interaction with the privatization dummy as additional controls. Column 3 further controls for GDP per capita, unemployment, income inequality, and public spending per capita. Column 4 adds a dummy variable for the political party that controlled the provincial government, and Column 5 includes the proportion of the population in the province with privatized water services. The pattern of the results is similar to that obtained for the nutrition pathway, in the sense that the coefficient of the lagged privatization dummy becomes not significant when standard errors are clustered at the province level.

Additional Evidence

Finally, survey data provided by the Ministry of Health are used in order to estimate the impact of the privatization of electricity services on households' probability of owning a refrigerator. The survey covers 30,000 households across the country and includes data on nutritional status. The survey identifies households with unmet basic needs, and household income is captured through a categorical variable that distinguishes whether the household is indigent, poor but not indigent, or not poor. This information is translated into two dummy variables, one for poor but not indigent households, and the other for non-poor households.

A probit regression is run for the probability of owning a refrigerator against a set of dummy variables: income, unmet basic needs, access to the electricity network, and province with a private electricity provider. Additional control variables include province GDP, the number of days with temperatures above 30 degrees Celsius, and income inequality. As reported in Table 3.7, the privatization dummy has a positive and significant effect. Its coefficient indicates that living in a province where electricity distribution has been privatized is associated, ceteris paribus, with an increase of about 2.2 percentage points in the probability of owning a refrigerator. Although just a correlation, and given that electricity access is controlled for, this result is consistent with the idea that privatization

Table 3.7	Impact of Privatization on the Probability of Having a Refrigerator

	Dependent Variable: Household with a Refrigerator (=1)
Private electricity service in the province	.0212
	(.0081)***
Temperature in the province	−.0001
	(.0001)
Ln(Real GDP per capita of the province)	.0284
	(.0089)***
Household with access to the electricity service	.4805
	(.0471)***
Income inequality	−.0039
	(.0021)*
Household with unmet basic needs	−.1687
	(.0058)***
Poor household (but not indigent)	.0493
	(.0051)***
Not-poor household	.1242
	(.0085)***
Observations	24,432

Notes: Standard errors clustered at the province level are in parentheses. Marginal effects are reported.
*Significant at the 10 percent level; **Significant at the 5 percent level; ***Significant at the 1 percent level.

led to an improvement in service quality, inducing households to buy refrigerators.

Conclusions

The central hypothesis of this chapter is that service expansions and quality improvements associated with the privatization of electric companies in Argentina have had a positive effect on health outcomes.

In order to test the main hypothesis, it is first shown that access to electricity service increased more in those provinces that privatized their electricity distribution networks than in provinces where distribution remained public. Also presented is evidence supporting the idea that private firms have better quality indicators than public firms.

The pathways whereby privatization increases access to service and quality of service are subsequently explored. First, by increasing access to electricity, privatization allows a number of households to have a refrigerator, which may improve nutrition intake. The empirical results show some evidence that in provinces where electricity distribution was privatized, the frequency of low birth weights (the measure of nutrition used in this chapter) decreases relative to provinces with public distribution networks, though the results are not robust to accounting for the potential problem of serial correlation in the data.

Second, by reducing the frequency and duration of interruptions, privatization may have an impact on the likelihood of food poisoning. As before, the empirical results show some evidence that provinces with privatized electricity systems have lower child mortality rates caused by food poisoning, though the results are not robust to correcting for correlation in the data.

The chapter also finds a positive and significant correlation between privatization and the probability of a household owning a refrigerator, a result that is consistent with the idea that privatization led to an improvement in service quality, inducing households to buy refrigerators.

The indirect benefits of electricity service privatization on health outcomes are not strong enough to provide policy implications beyond those implied by the results on access and service quality. The weakness of the results, however, might be a consequence of the low number of cross-section observations arising from working with province-level data in Argentina. Further research is needed in order to try to establish a causal effect of electricity privatization on health.

CHAPTER FOUR

Water Sector Privatization in Colombia

Felipe Barrera-Osorio and Mauricio Olivera

M anagement of the water sector is difficult. On one hand, water is essential for life, and access to good quality water is necessary in order to improve standards of living of the population. Moreover, externalities, such as effects on health, compel the state to intervene. On the other hand, private participation is likely to improve the efficiency of the sector in the provision of service. These two contradictory concerns can be resolved through private participation and a regulatory scheme that assures access to and quality of the service, especially for the poor. For that purpose, since the beginning of the 1990s Latin American countries have introduced various mechanisms of private participation (e.g., concessions, Build, Operate and Transfer—BOT—schemes, joint ventures, capitalization, and leases, among others) accompanied by regulation schemes. Colombia is no exception; its water sector has undergone two structural reforms since the end of the 1980s. First, water services were decentralized and now municipal governments are responsible for the provision of the service. Second, the 1991 constitutional reform introduced the mechanisms for private participation in this and other sectors that were controlled by the state.

Privatizations have lost momentum since then and have even en-countered significant political opposition. In 2002, according to Latino-barómetro, an annual survey in 17 Latin American countries, 61 percent of respondents disagreed with the statement that privatizations have been beneficial to the country.[1] In Colombia, 65 percent of the popula-

---

[1] 16,788 persons responded to this statement. See Carrera et al. (2004).

tion disagreed. The effect of water privatizations on welfare, however, has not been fully studied. In fact, cross-country evidence in the region is inconclusive. On one hand, Clarke, Kosec and Wallsten (2004) show in their study for Argentina, Bolivia and Brazil that access improved in both privatized and nonprivatized zones, suggesting that "Private Sector Participation, *per se*, may have not been responsible for these improvements" (Clarke, Kosec and Wallsten, 2004, p. 1). On the other hand, McKenzie and Mookherjee (2003) conclude in their study for Argentina, Bolivia, Mexico and Nicaragua that increased access at the bottom of the distribution outweighs the negative effect of increased prices. The effects on quality and on poverty, though, are not conclusive.

In order to address this uncertainty, this chapter studies the effects of water privatizations on consumer welfare in Colombia. In particular, the chapter measures the impact on access, quality and prices of water as well as on health outcomes. In spite of the unpopularity of privatizations, the main results suggest that privatizations have positive effects on welfare, especially in urban areas. In privatized urban municipalities, there is an improvement in access to water, and an increase in the quality of water, measured as the need for water treatment or as the aspect (clarity) of water. The frequency of the service, another measure of quality, decreases in privatized urban areas, but it increases for the lower quintiles. There is also a positive impact on health outcomes as measured by the weight for height ratio of children. In addition, in the urban areas of municipalities with better government technical capacities there are positive effects on access, payment and quality. In rural areas the negative effect of privatization on prices and strong negative effects on access to water outweigh the positive impacts on the improved frequency of the service and on improved child health, even after controlling for migration to these areas.

As the results show, the low levels of support for privatizations reflected in surveys like Latinobarómetro can be explained through price effects. However, as discussed below, privatization was undertaken simultaneously with the elimination of a cross-subsidy scheme, a factor that complicates explanation of changes in price. In addition, in terms of political economy the challenge is to expand the positive effects in urban areas to rural areas as well.

Water Privatization

The rules of the game for private participation in the utility sector in Colombia were introduced in the 1991 constitutional reform. While Colombia began the privatization process relatively late in comparison to other countries in the region, the country caught up rapidly during the period of 1994–1997. Three factors help to explain this performance. First, until the mid-1990s, Colombia was the most stable economy in the region, with high and stable growth, stable inflation, and a relatively low fiscal deficit. As a result, there was no urgent need to adjust public finances. Second, political factors, especially the role of labor unions, impeded the privatization of utilities. For example, the attempt to privatize the state-owned telecommunications company began in 1992 but was not achieved until 2004 due to labor union pressures. Third, since the beginning of the 1990s Colombia has decentralized the responsibility for most public utilities to subnational governments, and consequently the central government's power to influence privatization has become weak and indirect.

The share of private participation in the water sector is low compared to private participation in other sectors (i.e., 2 percent of total private participation). However, as in other countries, private participation in this sector has limitations for a variety of reasons. First, until the end of the 1980s, the sector was a state monopoly with lower potential competence and high fixed costs. Second, externalities in this sector are more pronounced. The most important cited example is health, but there are others such as environmental, gender, and poverty externalities. These externalities compel the state not to leave the sector completely in the hands of the private sector. In addition, the decentralization[2] of the sector could have affected private participation, as decentralization atomized the ownership of enterprises[3] and limited the capacity of the

---

[2] The decentralization process began in Colombia in 1986 with the popular election of mayors and was reinforced in the 1991 constitutional reform with the transfer of tax revenues and responsibilities.

[3] According to the Public Services Superintendency (2002), there are approximately 2,300 enterprises in the water and sanitation sector.

central government to influence the privatization process.[4] The central government can only intervene indirectly through financial and other incentives for subnational governments. The most important incentives implemented in Colombia are the Programa de Modernización Empresarial (PME) and the Programa de Privatizaciones y Concesiones en el Sector Agua, both designed in 1997. These programs give technical and financial assistance to municipalities to design private-public participation projects. Applying these incentives, the National Department of Planning (DNP) and the Ministry of Economic Development implemented three pilot projects in medium cities:[5] Montería, Pasto and Bucaramanga (DNP, 1997).

Table 4.1 shows private participation in the sector in municipalities with a population greater than 100,000, and Table 4.2 lists small city projects implemented by PME up to 2003. Some of the projects presented in the table were implemented before the design of the incentive programs (e.g., Barranquilla, Cartagena, Girardot, and Tunja). Others, like the Metroagua project in Santa Marta, were implemented later on but without the participation of the central government.

Two additional points are worth mentioning. First, given the sector's strong externalities, the government is in most cases responsible for financing long-term investment to assure coverage. Second, water and sanitation supply in the three largest Colombian cities—Bogotá, Medellín and Cali—are still managed by public enterprises.

In addition to the introduction of private participation in state-owned enterprises (SOEs) looking for economic efficiency, the 1991 constitutional reform introduced a regulatory framework to avoid perverse distributional effects and guarantee fair competition. As competition should benefit consumers, the regulatory framework intervenes through price mechanisms. These mechanisms include prohibiting prices lower than operational costs so that competitive enterprises are not driven out of business by other

---

[4] Until the end of the 1980s, this sector was managed by the Instituto de Fomento Municipal (Insfopal), a central government agency that was closed in 1986.

[5] Medium cities are cities with population greater than 100,000 and smaller than 1,000,000.

Table 4.1 Enterprises with Private Participation Serving Municipalities with Population Greater than 100,000

| Enterprise | Municipality | Private Participation | | | | Year of Privatization | Type of Contract | Partner |
		>50%	>50%, <10%	>10%	%			
Triple A*	Barranquilla	x			80	1991	Concession	Aguas de Barcelona
Acuacar*	Cartagena		x		45	1995	Concession	
CAMB-CDMB*	Bucaramanga			x	5		Joint venture	Local
Metroagua SA	Santa Marta	x			65	1996	Concession	Aguas de Barcelona
Aguas de Manizales	Manizales			x	1		Joint venture	Local
Acuaviva	Palmira	x			60		Concession	Suez
Proactiva	Monteria	x			100	2000	Concession	Proactiva
Aap	Popayan			x			Joint venture	
Conhydra	Buenaventura			x		2002		
Centro agua	Tulua	x						
Acuagyr	Girardot	x			70		Concession	Suez
Servaf	Florencia	x			51		Joint venture	Local
Aguas de Peninsula	Maicao	x				2001		
Sera q.a.	Tunja	x			100	1996	Concession	Proactiva
Aguas de Guajira	Rioacha	x				2000		
Sincelejo	Sincelejo					2002		
Soledad	Soledad					2002		

Source: Authors calculations from CRA (2002), DNP (2005), and Fernandez (2004).
Notes: * Serves cities with population greater than 600,000. The rest serves cities with population between 100,000 and 600,000.

Table 4.2	Projects with Private Participation Implemented by PME		
Department	Municipalities	Population	Initial Date of Operation
Atlantico	Sabanagrande y Santo Tomas (Asoasa)	44.000	August 2002
Atlantico	Ponedera	9.100	August 2002
Sucre	San Marcos	32.750	August 2002
Choco	Itsmina	13.500	October 2001
Choco	Trado	9.100	October 2001
Guajira	Barrancas, Distraccion, El Molino, Villanueva (Asoaguas)	42.700	June 2002
Cauca	Guapi	14.000	January 2002
Bolivar	San Juan Nepomuceno	27.000	December 2001
Huila	Nataga	1.800	April 2001
Magdalena	El Banco	51.700	February 2003
Meta	Cumaral	9.200	January 2002
Narino	El Charco	5.300	January 2002
Vichada	Puerto Carreno	7.500	January 2002

Souce: Fernandez (2004).

companies willing to take temporary losses in order to increase their market share; a second mechanism is the use of price caps to prevent excessive rent seeking by utilities (Prasad, 2005). The regulatory framework is also important for the water and sanitation sector since the increased openness of the economy introduced competition into all tradable markets, and the monopoly problem was confined to the public services that are often considered natural monopolies, such as energy distribution, water and sanitation services, and telecommunications.

Three laws govern the pursuit of these objectives: the Privatization Law (Law 226 of 1995), the public services law (Law 142 of 1994) and the electricity law (Law 143 of 1994). The Privatization Law additionally assures the participation of workers in the privatization process, or the democratization of property. For this purpose any SOE should be sold in two steps. In the first, the SOE is offered to its workers under special conditions in terms of length and loans, and with a fixed price. In the second, the SOE is offered to the public in general, with competitive price mechanisms. In general, the base price of the first step is lower than that in the second step (DNP, 1997).

The law of public services designed the regulatory framework. Among its most important characteristics, this law requires the incorporation of public service providers as "Sociedades por acciones" (ESP), which aims to isolate firm management from political intervention. Two other important features are related to the tariff regime. First, the law attempted to eliminate implicit subsidies, which are generally designed with political motives; at present there exists an explicit scheme of cross-subsidies to subsidize lower income strata. However, to assure providers' financial viability, making tariffs reflect their costs, the law set forth a transitional period in which this cross-subsidies scheme would be eliminated. Due to political pressure, the official deadline for closing the tariff gap between price and costs was extended until the end of 2005, although the cross-subsidy scheme nonetheless remains present in practice at the time of writing. In addition, given the intrinsic differences between the water and electricity sectors, the electricity law followed the general framework of the public services law and developed specific aspects of the links between generation, transmission and distribution of electricity with a long-run plan for expansion.

The public services law additionally defined the institutional organization of regulation. The regulatory scheme has a national scope, regulates both public and private enterprises, and is separated into two different agencies: a commission that is in charge of the policies of the sector, and a superintendency that is in charge of control. The three regulatory commissions for public services are the Energy and Gas Regulatory Commission (CREG), the Water and Sanitation Commission (CRA), and the Telecommunications Regulatory Commission (CRT). These commissions are responsible for planning the evolution of their sectors and guaranteeing the quality and coverage of public services. The Superintendencia de Servicios Públicos controls the performance of the public services sector and protects consumers' welfare.

After these reforms, the stylized facts presented below show that coverage and prices in the sector increased in both municipalities that privatized and municipalities that did not.[6] Water and sewerage cover-

---

[6] The stylized facts are based on DNP data, and on data collected by the Regulatory Commission and the Superintendency of Public Services.

age, measured by the number of users and by the number of households, increased in both privatized and nonprivatized municipalities (Table 4.3). In 2001, water coverage exceeded 90 percent in nonprivatized municipalities, and with the exception of Monteria, Santa Marta, Sincelejo and Florencia, access is also above 90 percent in privatized municipalities. Compared to these numbers, sanitation coverage is lower in both privatized and nonprivatized municipalities.

With respect to prices, there exists a cross price-subsidies scheme from the higher strata (5 and 6) to the lower strata (1, 2 and 3). However, following the transition toward eliminating the scheme defined by the law, tariffs in lower-income sectors have increased for municipalities both with and without private participation in the sector. Figures 4.1 to 4.4 show the change in tariffs per stratum between 2001 and 2004 for water and sewerage sectors in the four largest cities (Bogotá, Medellín, Cali and Barranquilla). In both sectors there has been an effort to reduce subsidies for the lowest strata, although tariffs in higher-income strata have also increased for sewerage.

Cross-subsidies are still high, and the increase in tariffs for the poor in order to close the gap and ensure the sector's long-run financial viability is important. Table 4.4 shows that on average the tariff for the poor is between 46 and 75 percent of the tariff of the higher strata, both for privatized and nonprivatized firms (on average between 60 and 65 percent, respectively). To close the gap between tariffs and costs, and to avoid the distortions generated by cross-subsidies, low-income households' tariff should increase between 35 and 40 percent in real terms, while for higher-income households the tariff should be reduced by approximately 13 percent.

As a result, water consumption has decreased. Table 4.5 shows the reduction in water consumption between 1997 and 2001 in the following main cities: Bogotá, Medellín, Cali, Barranquilla, Bucaramanga and Cartagena. As described before, in the former three cities water and sanitary services are provided by the public sector, while in the latter three they have been privatized. Reduction in consumption has been higher for low-income households, except in Barranquilla, and, to a lesser degree, Medellín. However, the tendency is similar in cities both with and without private participation. In sum, in the water and sanitation sector there is a trend

Table 4.3 Coverage of Water and Sewerage Systems

		Number of Users				Panel A: Municipalities with Private Participation — Coverage (%)						Annual Increase 1998–2001			
		Water		Sanitation		Water			Sanitation			Water		Sanitation	
Enterprise	Municipality	1998	2001	1998	2001	1990	1998	2001	1990	1998	2001	Users	Coverage	Users	Coverage
Triple A	Barranquilla*	226,918	253,116	175,339	209,778		89	94		76	80	3.7	1.7	6.2	1.7
Acuacar	Cartagena*	100,571	137,749	86,778	110,396		79	94		68	75	11.1	5.9	8.4	3.5
CAMB-CDMB	Bucaramanga*	161,771	171,365	160,465	169,606	84	99	99		99	99	1.9	0	1.9	-0.1
Metroagua SA	Santa Marta		61,833		51,092			85			70				
Aguas de Manizales	Manizales		77,126		73,401	82		96	77		93				
Acuaviva	Palmira	47,270	49,230	46,374	48,739		98	97		96	96	1.4	-0.4	1.7	-0.1
Proactiva	Monteria		39,746		19,646			76			38				
Aap	Popayan	44,554	49,275	41,779	46,606		95	98		90	93	3.4	1.1	3.7	1.1
Conhydra	Buenaventura														
Centro agua	Tulua	33,677	36,135	33,049	33,899		97	98		95	97	2.4	0.5	0.9	0.9
Acuagyr	Girardot		26,135		22,571			97			83				
Servaf	Florencia		22,509		12,285			85			46				
Aguas de	Maicao														
Sera q.a.	Tunja		27,777		27,296			98			96				
Aguas de Guajira	Rioacha														
EMPAS	Sincelejo		35,509		31,551			79			69				
	Soledad														

(continued on next page)

Table 4.3 Coverage of Water and Sewerage Systems *(continued)*

Panel B: Municipalities without Private Participation

Enterprise	Municipality	Number of Users — Water 1998	Water 2001	Number of Users — Sanitation 1998	Sanitation 2001	Coverage (%) — Water 1990	Water 1998	Water 2001	Coverage (%) — Sanitation 1990	Sanitation 1998	Sanitation 2001	Annual Increase 1998–2001 — Water Users	Water Coverage	Sanitation Users	Sanitation Coverage
EAAB	Bogotá*	1,195,147	1,206,160	1,082,045	1,206,160	88	93	95	64	93	95	0.3	0.7	3.7	0.6
EPM	Medellín*	694,485	760,821	680,939	728,853	98	99	100	94	94	93	3.1	0.3	2.3	-0.6
EMCALI	Cali*	422,899	436,799	402,295	418,677	89	92	96	75	94	94	1.1	1.4	1.3	0
E.I.C.E	Cucuta	106,869	121,991	93,913	118,489	86	90	96	88			8.1	-3	4.5	1.5
A y A	Pereira	89,934	87,193			77	94			82	96				
IBAL	Ibague	84,276	80,042							89	85				
EMPOPASTO	Pasto	53,804	53,529							89	89				
EE PP	Neiva	65,595	62,509			81	95			86	99				
EE PP	Armenia	67,543	66,874							96	95				
EAAV	Villavicencio	48,475	53,547	53,610	59,576	72	80	80	89			3.6	3.6	3.4	3.4
EMDUPAR	Valledupar	43,606	48,648	43,610	46,754	99	99	87	90			2.3	1.3	3.7	0
	Sogamoso	24,078	26,602	20,004	22,455	97	100	66	90			3.9	10.9	3.4	1.1

Source: Authors' construction from CRA (2002), DNP (2005), and Fernandez (2004).
Notes: * Serves cities with population greater than 600,000. The rest serves cities with population between 100,000 and 600,000.

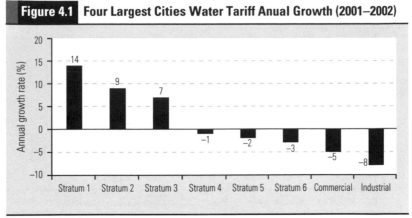

Figure 4.1 Four Largest Cities Water Tariff Anual Growth (2001–2002)

Source: SSPD (2005).

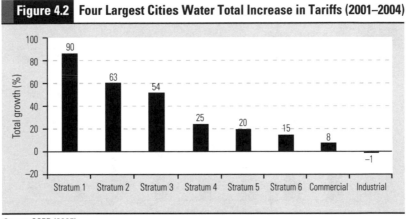

Figure 4.2 Four Largest Cities Water Total Increase in Tariffs (2001–2004)

Source: SSPD (2005).

of increasing coverage, higher prices and lower consumption, regardless of whether a municipalities' services are public or privatized.

Hypotheses and Econometric Strategy

The aggregate data from the previous section showed a similar pattern in prices and access for both privatized and nonprivatized municipalities. However, this is a restricted, partial analysis, since its source of variation

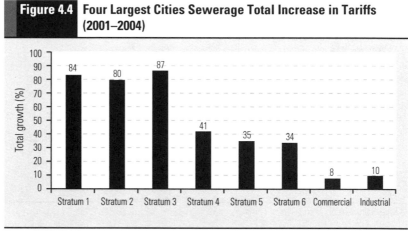

Figure 4.3 Four Largest Cities Sewerage Tariff Annual Growth (2001–2004)

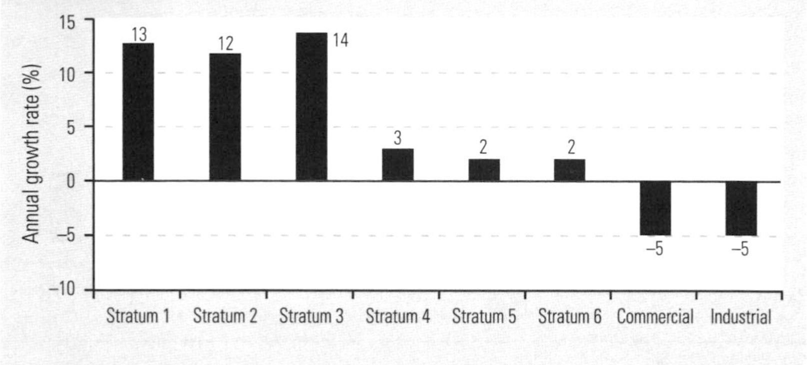

Source: SSPD (2005).

Figure 4.4 Four Largest Cities Sewerage Total Increase in Tariffs (2001–2004)

Source: SSPD (2005).

is at the municipal level and not from micro household data. This section presents the econometric strategy for studying the impacts of privatizations in the water and sanitation sector on access, payment, quality, and health, using two sources of micro data.

In general, a functional form is estimated of the following equation:

$$Y_{i,j,t} = f(X_{i,t}, X_{j,t}, D, \varepsilon_{i,j,t}) \tag{1}$$

Table 4.4	Average Tariff: (Pesos per m³/Month/Users in December 2001)

Panel A: Municipalities with Private Participation

Enterprise	Municipality	Lower Income/ Higher Income Tariff (%) **
Triple A*	Barranquilla	55.3
Acuacar*	Cartagena	58.5
Metroagua SA	Santa Marta	68.6
Aguas de Manizales	Manizales	70.0
Acuaviva	Palmira	75.4
Aap	Popayan	66.3
Acuagyr	Girardot	67.9
Servaf	Florencia	65.1
	Average	65.9

Panel B: Municipalities without Private Participation

EAAB*	Bogotá	58.3
EPM*	Medellín	55.5
EMCALI*	Cali	66.6
E.I.C.E	Cucuta	68.3
A y A	Pereira	54.2
IBAL	Ibague	65.6
EMPOPASTO	Pasto	60.6
EE PP	Neiva	60.2
EE PP	Armenia	46.0
EAAV	Villavicencio	71.7
EMDUPAR	Valledupar	54.2
EMPAS	Sincelejo	62.1
	Average	59.3

Source: Authors' calculation based on SSPD (2002)
Notes: * Serves cities with population greater than 600,000. The rest serves cities with population between 100,000 and 600,000.
** Lower income is the average for strata 1, 2 and 3; higher income is the average for strata 5 and 6.

where the impact variable (denoted as $Y_{i,j,t}$) may be access to water, prices, quality of the service, or health outcomes for individual i living in municipality j at time t; D is a dichotomous variable equal to 1 for individuals in municipalities in which privatization took place and equal to 0 for the comparison group of individuals (i.e., individuals in municipalities without privatization); $X_{i,t}$ is a vector that includes control variables of the individuals and $X_{j,t}$ a vector of municipal characteristics that affect the impact variable; and finally $\varepsilon_{i,j,t}$ captures the unobservable characteristics

| Table 4.5 | Average Consumption (m³/Month/User) | | | | | | | |

	Bogotá				Medellín			
	1997	2001	Total Variation	Annual Variation	1997	2001	Total Variation	Annual Variation
Stratum 1	21.4	14.8	−30.9	−8.8	17.9	13.8	−22.9	−6.3
Stratum 2	22.7	15.4	−32.2	−9.3	19.7	14.9	−24.2	−6.8
Stratum 3	22.0	14.5	−34.2	−9.9	19.6	16.0	−17.4	−5.0
Stratum 4	17.7	13.9	−21.5	−5.9	20.8	18.0	−14.0	−3.6
Stratum 5	19.5	16.2	−17.1	−4.6	24.7	20.2	−17.3	−4.8
Stratum 6	21.8	19.0	−13.1	−3.4	36.0	26.8	−25.5	−7.1
	Cali				Barranquilla			
Stratum 1	23.8	20.0	−16.1	−4.3	14.0	15.7	−11.8	−2.8
Stratum 2	27.6	21.9	−20.6	−5.6	31.1	23.1	−25.8	−7.2
Stratum 3	23.8	21.3	−10.6	−2.8	35.7	23.2	−35.0	−10.2
Stratum 4	23.6	21.6	−8.2	−2.1	37.9	23.8	−37.2	−11.0
Stratum 5	27.6	25.9	−6.3	−1.6	43.2	26.2	−39.4	−11.8
Stratum 6	35.2	35.6	−0.9	0.0	47.6	32.8	−31.1	−8.9
	Bucaramanga				Cartagena			
Stratum 1	29.3	18.9	−35.7	−10.4	21.6	12.8	−40.9	−12.3
Stratum 2	28.4	19.9	−30.0	−8.5	24.2	15.7	−35.0	−10.2
Stratum 3	25.2	19.1	−24.1	−6.7	25.9	17.6	−32.1	−9.2
Stratum 4	24.6	21.4	−13.0	−3.4	27.0	20.0	−25.7	−7.2
Stratum 5	27.8	24.8	−10.8	−2.8	29.9	23.3	−21.8	−6.0
Stratum 6	31.9	29.7	−6.9	−1.8	33.2	22.7	−31.8	−9.1

Source: Authors' calculations based on SSPD (2002).

of the individuals. The subscript t denotes time variation (i.e., 1997, before privatization, and 2003, after privatization).

A family in a difference-in-differences model is estimated under one identification assumption. Variation at the household and municipality levels, time variation (i.e., before and after privatization), and "treatment" variation (i.e., privatized and nonprivatized municipality) are exploited.

The specification of Equation (1) to be estimated is the following:

$$Y_{i,j,t} = \beta_0 + \beta_1(t*D) + \beta_2 D + \beta_3 t + B_4 X_{i,t} + B_4 X_{j,t} + \varepsilon_{i,j,t} \qquad (2a)$$

and four estimations of the same equation are presented: first, without controlling for the indices of technical capacity for the municipalities, but controlling for fixed effects at municipalities; second, controlling for the indices of technical capacity (invariant over time); third, estimating for urban and rural areas separately; and finally, including the possibility of differential effects across quintiles of income. The last variation is estimated using the following equation:

$$Y_{i,j,t} = \beta_0 + \beta_1(t*D) + \beta_2 D + \beta_3 t + B_4 X_{i,t} + B_4 X_{j,t} + \beta_5 Q_{i,t} + \qquad (2b)$$
$$\sum_k \beta_{6,k}(Q_{i,t,k}*D*t) + \varepsilon_{i,j,t}$$

where Q represents the quintiles. The impact of privatization is given by the difference-in-difference estimator (DD):

$$DD = \beta_1 D + \beta_2[(X_{t=1}^T - X_{t=0}^T) - (X_{t=1}^C - X_{t=0}^C)] + \qquad (3)$$
$$[(\varepsilon_{t=1}^T - \varepsilon_{t=0}^T) - (\varepsilon_{t=1}^C - \varepsilon_{t=0}^C)]$$

which gives the effect of the treatment ($\beta_1 D$).

Endogeneity problems in estimations of Equations (2a) and (2b)[7] occur when there are characteristics that are not observable by the researcher but influence the decision to carry out the privatization. For instance, suppose that privatization occurs in municipalities in which technical capacity is higher than in the rest of the country, presumably because the human capital of that municipality is higher. Also, suppose that the dependent variable $Y_{i,j,t}$ is infant mortality, which presumably depends not only on the quality of the water but also on the "quality" of the household. Since a "better" quality of household and higher technical capacity are difficult to quantify, these unobservable characteristics will be included in the term $\varepsilon_{i,j,t}$ in Equation (3). At the same time, individuals in these municipalities may have a greater probability of living in a municipality in which privatization occurred and, therefore, $E(D, \varepsilon_{i,j,t}) \neq 0$ which makes β_i a biased estimate.

To tackle the problem of endogeneity, the following identification assumption is used. The changes in the unobservable characteristics

---

[7] For a general discussion of endogeneity problems and econometric strategies, see Heckman, Lalonde and Smith (1999).

are equal across groups (treatment and control) or—a stronger assumption—the unobservable characteristics are equal for the treatment and control groups at each period of time. Hence, to estimate the true impact of privatization, the following identification assumption is used: the unobservable characteristics are time invariant.

In order to see why this makes the DD unbiased, assume that $\varepsilon_{i,j,t}$ presents the structure $\varepsilon_{i,j,t} = \eta_i + \mu_{i,j,t}$, where η_i is the unobservable set of characteristics that are time invariant and $\mu_{i,j,t}$ is a pure random component. In this case, $(\varepsilon_{t=1}^i - \varepsilon_{t=0}^i) = (\eta_i - \eta_i) - (\mu_{t=1}^i - \mu_{t=0}^i) = (\mu_{t=1}^i - \mu_{t=0}^i)$ is random and not related to the fact that the person is under the effects of privatization, and therefore, $DD = \beta_i$.

Data

For the dependent ($Y_{i,j,t}$) variables information is taken from the Encuesta de Calidad de Vida (ECV) for the years 1997 and 2003. These household surveys have an extended questionnaire, with chapters that include information on household education, health, income, consumption, and shocks, as well as the characteristics of the household's dwelling unit. These data are complemented with health variables, related specifically to diseases caused by poorly treated water, from the 1995 and 2005 Demographic and Health Surveys (DHS).[8]

It would be ideal to use a panel of *households* to compare outcomes through time, but unfortunately the available data do not permit that approach. Instead, a panel is constructed of *municipalities* for which information is available in both years. Table 4.6 presents the municipalities for which there are observations for both years in the ECV survey (1997 and 2003), and for the two health variables, diarrhea and weight for height, from the 1995 and 2005 DHS. Based on the privatization process discussed above, municipalities where water privatization took place are identified and an unbalanced panel dataset is formed at the municipality level. There are a total of 46 municipalities with observations in both years

---

[8] These surveys are available at www.measuredhs.com.

Table 4.6 Municipalities that Appear in Both Years of the Survey

ECV			Additional from DHS		
Municipality	Privatized	Non-privatized	Municipality	Privatized	Non-privatized
Aratoca		x	Alto Baudó		x
Arauca		x	Baranoa		x
Barbosa		x	Betulia		x
Barrancabermeja		x	Buenaventura	x	
Barranquilla	x		Buga		x
Bello		x	Cajibío		x
Bogota		x	Candelaria		x
Bucaramanga	x		Caparrapí		x
Caicedo		x	Chigorodó		x
Caldas		x	Chinú		x
Cali		x	Corozal	x	
Cartagena	x		Coyaima		x
Chia		x	Cumbal		x
Cienaga		x	El Guacamayo		x
Copacabana		x	Istmina	x	
Cucuta		x	Majagual		x
Dos Quebradas		x	Mogotes		x
Envigado		x	Neiva		x
Floridablanca		x	Ortega		x
Girardota		x	Palmira	x	
Giron		x	Plato		x
Ibague		x	Popayán	x	
Itagui		x	Puerto Salgar		x
Lorica		x	Riosucio		x
Los Patios		x	Saboya		x
Manizales	x		Salgar		x
Medellín		x	San Bernardo del Viento		x
Montenegro		x	San Jacinto		x
Monteria	x		San Martín		x
Pasto		x	Sibaté		x
Pereira		x	Sogamoso		x
Piedecuesta		x	Subachoque		x
Pitalito		x	Trujillo		x
Providencia		x	Túquerres		x
Riohacha	x		Turmequé		x
Roldanillo		x	Urumita		x
San Andrés		x			
Santa Marta	x				
Santander de Quilichao		x			
Soledad	x				
Tulua	x				
Tumaco		x			
Villa del Rosario		x			
Villamaria		x			
Yopal		x			
Yumbo		x			

for the ECV, and 36 more (for a total of 82) municipalities with health variables coming from the DHS for Colombia.

Water privatizations presumably have an impact on the coverage, price, and quality of water, as well as on the incidence of health problems related to poor quality water. Coverage is measured as a 0/1 dummy capturing access to water (e.g., connected to water services or not), price as the monthly payments made by the household (as percentage of total income[9]) for water consumption, quality as the need for treatment for consumption, the frequency of service (continuity of services through the week) and the aspect (if it is turbid, or with particles) of the service, and for health variables we include children with diarrhea in the last two weeks (a 0/1 dummy) and a measure of weight for height. As the literature shows (for instance, Galiani, Gertler and Schargrodsky, 2005), health indicators are expected to be particularly affected by the provision of water.

Table 4.7 presents the means of the dependent measures for privatized and nonprivatized municipalities in 1997 and 2003. Access to water is relatively high, with 94 percent and 96 percent of households, respectively, having water service. Such high coverage may imply that marginal increments are quite difficult to achieve. Coverage does, however, vary between urban and rural areas, and to this end both the data and the estimations are presented, allowing for heterogeneity responses by urban and rural areas. Coverage of service increases with time, with no statistical differences between privatized and nonprivatized municipalities. Expenditure on water services, as a percentage of the total income of the household, increases over time in both privatized and nonprivatized municipalities. In terms of the quality of the service, the indicators show overall improvements between 1997 and 2003 in both types of municipalities. Households do not report a difference in the need to treat piped water. However, with respect to the frequency and the aspect of water, privatized municipalities present worse outcomes than nonprivatized ones. Regarding health outcomes from the ECV, in both privatized and nonprivatized municipalities a general improvement is observed between

---

[9] This dependent variable was calculated also as monthly payments in 2003 pesos. The results do not change significantly.

Table 4.7	Descriptive Data

	Panel Municipalities		Privatized		Nonprivatized		Difference in Difference
	1997	2003	1997	2003	1997	2003	
Outcome Variables							
a. Coverage							
Access (Dummy 1–0)	0.940	0.959	0.910	0.936	0.948	0.966	−0.001
Payment	0.076	0.086	0.081	0.095	0.075	0.084	0.005
b. Quality							
Water treatment (Dummy 1–0)	0.318	0.507	0.330	0.465	0.315	0.519	−0.069
Frequency of Service (Dummy 1–0)	0.800	0.859	0.771	0.806	0.807	0.874	−0.031*
Aspect (Dummy 1–0)		0.844		0.819		0.851	−0.032*
c. Health outcomes							
Health status (Dummy 1–0)	0.724	0.772	0.719	0.768	0.725	0.773	−0.026
Hospitalizations (Dummy 1–0)	0.074	0.068	0.061	0.072	0.077	0.068	0.019
Health problems (Dummy 1–0)	0.158	0.102	0.149	0.155	0.160	0.098	0.053
Diarrhea (Dummy 1–0)[1]	0.157	0.133	0.177	0.145	0.151	0.131	0.007
Height for weight[1]	0.095	0.060	−0.068	−0.100	0.138	0.090	0.005
Household Variables							
a. Infrastructure							
Zone (Urban or rural)	1.119	1.085	1.132	1.086	1.116	1.084	−0.015
Ownership (Dummy 1–0)	0.806	0.869	0.778	0.809	0.813	0.886	−0.043*
Rooms (Number of)	3.506	3.504	3.542	3.547	3.497	3.492	0.011
Risk (Dummy 1–0)	0.001	0.124	0.000	0.169	0.001	0.110	0.060*
Infrastructure index	0.688	0.669	0.645	0.610	0.700	0.686	−0.021*
Assets index	0.476	0.482	0.457	0.459	0.480	0.488	−0.006
b. Human capital							
Number of persons per house	4.022	3.699	4.278	3.952	3.955	3.627	0.002
Head's education (Total years)	7.473	8.033	7.405	8.075	7.490	8.021	0.140
Average education (Total years)	7.841	8.895	7.841	9.026	7.841	8.858	0.168
Head's age	46.200	47.009	45.686	45.586	46.334	47.415	−1.180*
Recent migrants	0.078	0.100	0.082	0.109	0.077	0.098	0.005
Head's marital status (Dummy 1–0)	0.676	0.631	0.688	0.634	0.672	0.630	−0.011
Head's gender (Dummy 1–2, 1=Male)	1.279	1.336	1.275	1.340	1.280	1.335	0.010

(continued on next page)

Table 4.7 Descriptive Data (*continued*)

	Panel Municipalities		Privatized		Nonprivatized		Difference in Difference
	1997	2003	1997	2003	1997	2003	
c. Income							
Per capita income (Constant $2003)	$744,644	$466,381	$497,036	$404,670	$809,002	$483,980	$232,656*
Per capita consumption (Constant $2003)	$569,071	$538,631	$463,272	$490,800	$596,570	$552,272	$71,825*
Head's labor income (Constant $2003)	$1,000,131	$647,908	$874,614	$534,086	$1,033,791	$680,893	$12,371
Head employed	0.758	0.738	0.764	0.750	0.757	0.734	0.009
Municipalities Variables							
a. Exogenous characteristics							
Recent migration (Dummy 1–0)	0.105	0.125	0.085	0.091	0.082	0.133	−0.018
NBI	26.610		29.690		25.744		−3.946
Population	441576.7		452813.8		438416.3		−14397.5
Surface area (km²)	966.250		1409.333		852.314		−557.019
Distance to state capital (km)	49.818		12.222		59.486		47.263
Altitude (from sea level)	1016.886		480.889		1154.714		673.825
b. Technical capacity indices (TCI)							
Own resources	23.828	26.740	25.772	24.573	23.282	27.267	−5.184
Saving capacity	57.357	74.345	56.354	77.725	57.639	73.523	5.487
Income from national transfers	40.060	54.337	43.758	63.757	39.021	52.046	6.973

Source: Authors calculations with data from DANE for municipalities characteristics, DNP for TCI.
Note: *Statistically significant at the 5% level; **Statistically significant at the 10% level; Difference in time and between privatized and nonprivatized.
[1] These two variables come from the DHS survey and the years before and after privatizations are 1995 and 2005, respectively.

1997 and 2003. Similarly, the diarrhea and height for weight indicators from the DHS present improvements between 1995 and 2005. There are no statistical differences in these indicators for privatized and nonprivatized municipalities over time.

Two types of control variables are used, one for household characteristics $(X_{i,t})$ and one for municipal characteristics $(X_{j,t})$. The sources for the first type of controls are the Encuesta de Calidad de Vida or the

DHS, depending on the impact estimated.[10] The sources for the second type of controls are the National Department of Planning (DNP) and the National Department of Statistics (DANE).

At the household level, measures of housing characteristics, human capital and income are used. Housing characteristics include the following measures: whether the dwelling unit is in a rural or urban area, the type of ownership, and indices capturing the infrastructure of the house and the assets of the house. The infrastructure index includes the type of floor (0 if it is inadequate, 0.5 if it is fairly adequate and 1 otherwise), a natural risk factor (0 if the household faces a natural risk, 1 if not) and number of individuals per bedroom (0 if there are two or more people per bedroom, 1 otherwise). The asset index includes the existence of a washing machine, refrigerator, blender, oven, motorcycle, and car. Both indexes go from 0 to 1, where 1 indicates the highest quality of infrastructure or highest possible asset accumulation and 0 the lowest. Human capital variables include years of education (of all adults and of the head of the household) and some demographic variables (such as the size of the household, marital status and gender of the head, among others). Finally, income variables include income of the household from all sources (in constant pesos of 2003), total consumption (in constant pesos of 2003), and employment, measured via occupancy (yes or no) of the head or all the adults in the household.

Housing characteristics show the following trend. The percentage of rural households in the sample is between 11.9 percent and 8.5 percent for the 1997 and 2003 samples, respectively. The 2003 sample tends to be more urban because of over-sampling of Bogota in 2003, and there are no differences in the percentage of households in rural areas for the privatized and nonprivatized municipalities. By 2003 around 87 percent of households owned a house (including houses that were totally paid for and houses whose owners were still making payments). Household ownership increased between 1997 and 2003, at a faster rate for the

---

[10] The same household controls as in the DHS are used here, except a measure of income. As shown below, the unavailability of these data prevented an estimation of the effects of privatizations on health by quintiles.

nonprivatized municipalities. On average, the infrastructure index presented a small reduction over time. A typical house presents 60 percent of adequate infrastructure, and the index decreased slightly more for the privatized municipalities than for the nonprivatized ones. In terms of the asset index, the typical household had in both years around 50 percent of the assets included in the index. There are no differences across time or across types of municipalities.

Human capital characteristics are not different between privatized and nonprivatized municipalities, with the exception of the age of the head of household, which is younger in privatized municipalities. The sample additionally shows improvement of indicators for number of households and education, and there is some trend towards more migration in 2003 than in 1997.

Average income falls dramatically between 1997 and 2003. In 1999 the country suffered a severe depression, and the recovery was very slow. This decline was especially pronounced for urban as opposed to rural areas, and for nonprivatized as opposed to privatized municipalities; the latter result is driven by Bogota. In terms of employment, there is no difference in the employment rate of household heads, which is around 74 percent. In general, there are no systematic differences between privatized and nonprivatized municipalities regarding the "exogenous" variables. Some exceptions are per-capita income and homeownership. Despite these cases, there was not a clear direction of the differences.

The second set of control variables are those that vary at the municipal level. The first set of municipal variables is fixed and time-invariant, and there are no differences between privatized and nonprivatized in any of the municipal variables. In particular, there are no differences between the two groups in terms of surface area, distance to the state capital and altitude. Similarly, there are no differences in terms of the level of Unmet Basic Needs in 1993, and both types of municipalities had the same starting point in terms of income.

The similarity in exogenous variables at both household and municipal levels between privatized and nonprivatized localities is critical in order to isolate the true effects of the privatizations from other, exogenous differences in characteristics between privatized and nonprivatized municipalities, thus ensuring that $(X^T_{t=1} - X^T_{t=0}) = (X^C_{t=1} - X^C_{t=0})$ will reduce the source of bias.

The second set of municipal variables includes fiscal indicators that can be divided into three groups: indicators of technical capacity, municipal variables exogenous to the privatization process, and fiscal variables that may be endogenous to the privatization process. Indicators proxy to technical capacity are an important component of the effect of privatizations on welfare in the sense that there is a branch of the literature that proposes the hypothesis that privatization "works" in municipalities that have enough technical capacity to take advantage of the process. However, another type of causality may emerge. Indeed, technically sound municipalities that have been successful in running public services are less likely to privatize compared to municipalities with lower technical capacity that are unsuccessful in the provision of services. The argument is developed in the discussion of the results.

The proxies chosen for technical capacity include income from local taxes as percentages of total municipal revenues (Local taxes revenues/Total revenues) and savings capacity ((Current Income-Current Expenditure)/Current Income), borrowing several ideas from the public finance literature (for the general case, see Bird and Smart, 2002, and Fiszbein, 1997; and Chaparro, Smart and Zapata, 2006, for the case of Colombia). On one hand, local income taxes depend on three variables: the tax base, the level of the tax and the technical capacity of the municipality. Once local income is normalized dividing by total income, two municipalities that have an equal tax base and tax rate will have differences in the ratio because of the technical capacity for collection. In turn, technical capacity will depend on the human capital and the administrative capacity of the locality (Fiszbein, 1997). One clear example of this proxy of technical capacity is that local revenues depend, for instance, on the ability of the municipality to create a good measure of the size of property within the municipality. On the other hand, savings capacity is a critical variable for fiscal purposes. Local investment depends on the amount of transfers, debt and savings capacity. Usually, more advanced and sound municipalities will use a bigger proportion of savings capacity to finance its own investment. For this reason, savings capacity can serve as well as a measure of technical capacity of the municipality. In short, municipalities with higher administrative and technical capacity are able to increase their local tax base, and municipalities with greater administrative capacity are able to accumulate more current savings. These proxies are included in order to investigate the hypothesis of differential impacts of privatization depending on technical and administrative capacity, and taking into account that the regulatory framework applies to all firms, public and private, due to its centralization.

The group of fiscal variables exogenous to the privatization process includes transfers from central government revenues. These transfers are determined by a set of rules outside the municipalities' decision-making power and therefore in the short run exogenous to the municipality. Finally, the group of fiscal variables endogenous to the privatization process includes current income earmarked for administrative expenditures and investment expenditures as a percentage of total expenditure. Clearly, privatization can ease the budget constraint and induce a reallocation of expenditure. Nevertheless, these two variables were not used as controls in the estimation, since they could generate endogeneity problems.

In general terms, there exists an apparent upward time trend in revenues from local taxes for nonprivatized municipalities.[11] For privatized municipalities, there is a drastic drop between 1994 and 1995, and stability since then. Despite the apparent difference between privatized and nonprivatized municipalities, there is no significant difference at a 90 percent confidence interval. Current savings present a clear upward trend, from 1994 to 2004, for both privatized and nonprivatized municipalities with no significant difference between the privatized and nonprivatized municipalities. In sum, there is no evidence of differences in technical capacity between one group and the other.

Likewise, there appears to be no difference between the respective transfers to privatized and nonprivatized municipalities. Transfers have been

---

[11] The data cited in this paragraph and the following paragraph are drawn from the figures in Appendix 1 of Barrera-Osorio and Olivera (2007), the working paper on which this chapter is based.

increasing steadily since 1994. Current income intended for current operating expenses presents a clear negative slope since 1994, with no significant difference between privatized and nonprivatized municipalities. Finally, revenues used for investment have been increasing since 1994 for both types of localities, again with no statistical difference between types.

Results

Tables 4.8 to 4.12 present the results of the estimations of Equations (2a) and (2b). For all the impacts the tables report the coefficients for the dummy that captures time changes [denoted by Time (t)], for the dummy that captures if the municipality is privatized or not [Private (P)], and for the impact ($P{*}t$, which corresponds to the DD estimator). All regressions include household and municipal controls, and regressions are reported

Table 4.8	Water coverage						
Dependent Variable: Access (Dummy 1–0)							
			Quintiles	Urban		Rural	
Privatized (P)	−0.0057	0.0042***	0.0028*	0.0077**	−0.1035	0.0353	0.5103***
	−0.0068	−0.0015	−0.0016	−0.0032	−0.1535	−0.1561	−0.1537
Time (t)	−0.0008	−0.0014**	−0.0005	−0.0011	−0.0040**	−0.0193	0.0624
	−0.0005	−0.0006	−0.0008	−0.0008	−0.0017	−0.0332	−0.0570
P*t	−0.0003	−0.0031	0.0024***	0.0012	−0.4570***	−0.5746***	
	−0.0012	−0.0020	−0.0007	−0.0008	−0.0589	−0.0570	
Quintile 1*P*t			−0.0022				
			−0.0032				
Quintile 2*P*t			−0.0046				
			−0.0046				
Quintile 3*P*t			−0.0025				
			−0.0038				
Quintile 4*P*t			−0.0149				
			−0.0215				
Quintile 5*P*t			−0.0007				
			−0.0043				
R²	0.58	0.58	0.59	0.35	0.37	0.28	0.31
No. Obs.	21406	21003	21003	15852	15693	2197	1922
TCI	No	Yes	Yes	No	Yes	No	Yes

Notes: Standard errors in parenthesis.
All regressions include the controls described in the text and in Table 4.7.
* significant at 10%, ** significant at 5%, *** significant at 1%

both with and without variables that control for technical municipal capacity. The tables also show models that permit heterogeneity of treatment across quintiles of income, and for urban-rural subsamples.

Briefly, the rationale for including these four types of regressions (without controlling for technical capacity, controlling for technical capacity, quintile heterogeneity and urban/rural heterogeneity) is the following. First, controlling (or not) for technical capacity is important in light of the privatization literature emphasizing that privatization is more successful in municipalities with higher technical capacity. Second, allowing for quintile heterogeneity is important in order to determine the effects of privatization on households at the left of the income distribution. Finally, allowing for urban/rural differences is important because, as shown above, average levels of connection to water are very high in Colombia. As it turns out, the effects of privatization are *greater* in urban than rural areas; privatization apparently works best in high-density areas that probably had infrastructure in place before privatization.

For access, there is no evidence of positive impact of privatization when the model does not include municipalities' technical capacity. The results do not show statistical differences between privatized and nonprivatized localities, even after controlling for technical capacity. In addition, there are no significant differences in the effects of privatization on access across income quintiles. However, when urban and rural subsamples are estimated, coverage increases in privatized urban areas by around 2.4 percentage points. This effect disappears when municipal technical capacity is included as a control, suggesting that municipalities with technical capacity can somehow "compete" with private companies; an example of this occurs in Medellín, where the Medellín Public Enterprise (EPM) is one of the most efficient public enterprises. In rural areas, however, there is a strong negative effect on access in rural areas, as shown in Table 4.8.

On average, no effects of privatization on prices are observed. However, when heterogeneous responses from different income quintiles are allowed, the data show higher price increases for lower quintiles. Two reasons explain this result. First, due to the elimination of cross-subsidies in the water sector, quintiles 1 and 2 spend more of their total income for this utility, while the fourth quintile pays less. The effect is important. Lower quintiles pay an additional 7 to 11 percent in relation to their income, while for the fourth quintile the share of income used to pay for water declines by 10 percent. Second, households in rural areas that privatized spend a higher amount of their income on water and sewerage, an estimated 3.2 percent of their income. This effect disappears, however, when technical capacity is controlled for (Table 4.9).

With respect to quality, households in localities that privatized—particularly households in urban privatized areas—have to engage in less additional treatment of water than in nonprivatized localities, independently of the technical capacity of the municipality (Table 4.10, Panel A). When the quality of water is measured as the frequency of the service, it is higher in nonprivatized urban municipalities, although the second quintile presents an improvement in the frequency of the service compared to

Table 4.9 **Payment**

Dependent Variable: Payment (% of total income of the household)							
			Quintiles	Urban		Rural	
Privatized (P)	−0.0422	−0.0134	−0.1113	−0.0246	−0.0192	−0.0273*	−0.0205
	−0.0737	−0.2463	0.2466	−0.0236	−0.0316	−0.0146	−0.0218
Time (t)	0.0103**	0.0203**	0.0043	0.0106	0.0238	0.0142**	0.0140
	−0.0046	−0.0080	0.0176	−0.0066	−0.0170	−0.0057	−0.0091
P*t	−0.0076	−0.0021	−0.0092	−0.0034	−0.0085	0.0318*	
	−0.0105	−0.0116	−0.0328	−0.0337	−0.0104	−0.0177	
Quintile 1*P*t			0.0713**				
			0.0359				
Quintile 2*P*t			0.1116***				
			−0.0266				
Quintile 3*P*t			−0.0140				
			−0.0236				
Quintile 4*P*t			−0.1029***				
			−0.0215				
Quintile 5*P*t			−0.0037				
			−0.0204				
R²	0.04	0.04	0.05	0.04	0.04	0.20	0.20
No. Obs.	16398	16168	16168	15321	15201	1077	967
TCI	No	Yes	Yes	No	Yes	No	Yes

Notes: Standard errors in parenthesis.
All regressions include the controls described in the text and in Table 4.7.
* significant at 10%, ** significant at 5%, *** significant at 1%

Table 4.10 Quality

Dependent Variable: Water Treatment (Dummy 1–0)						
		Quintiles	Urban		Rural	
Privatized (P)	−0.3349***	−0.3045*	−0.2561	−0.3632***	0.0045	0.2504	−0.1285
	−0.0646	−0.1684	−0.1825	−0.0860	−0.1438	−0.1962	−0.2065
Time (t)	0.2401***	0.2961***	0.1643***	0.2620***	0.3407***	0.1085***	0.0628
	−0.0121	−0.0194	−0.0369	−0.0132	−0.0227	−0.0305	−0.0478
P*t	−0.1011***	−0.0868***	−0.1021***	−0.0644*	0.0286	0.0981	
	−0.0276	−0.0299	−0.0310	−0.0354	−0.0662	−0.0754	
Quintile 1*P*t			0.0425				
			−0.0764				
Quintile 2*P*t			−0.0474				
			−0.0618				
Quintile 3*P*t			−0.0800				
			−0.0624				
Quintile 4*P*t			−0.1324**				
			−0.0594				
Quintile 5*P*t			−0.0342				
			−0.0660				
R²	0.1494	0.1516	0.1617	0.1692	0.1717	0.1251	0.1242
No. Obs.	22475	22072	22072	20076	19888	2383	2168
TCI	No	Yes	Yes	No	Yes	No	Yes
Dependent Variable: Frequency of Service (Dummy 1–0)						
Privatized (P)	0.2784***	0.2963***	0.2570***	0.2547***	0.2719***	−0.5069**	−0.6777***
	−0.0246	−0.0287	−0.0316	−0.0263	−0.0312	−0.2161	−0.1917
Time (t)	0.0587***	0.0564***	0.0384	0.0692***	0.0489***	−0.0884**	−0.1134
	−0.0090	−0.0136	−0.0246	−0.0094	−0.0165	−0.0413	−0.0758
P*t	−0.0381*	−0.0421*		−0.0544**	−0.0306	0.2158***	0.2357***
	−0.0217	−0.0220	−0.0240	−0.0219	−0.0655	−0.0699	
Quintile 1*P*t			0.0184				
			−0.0384				
Quintile 2*P*t			0.0614***				
			−0.0173				
Quintile 3*P*t			0.0091				
			−0.0329				
Quintile 4*P*t			−0.1999***				
			−0.0733				
Quintile 5*P*t			−0.2831***				
			−0.0849				
R²	0.3178	0.3087	0.3180	0.3461	0.3275	0.2247	0.2190
No. Obs.	20570	20296	20296	18975	18796	1311	1225
TCI	No	Yes	Yes	No	Yes	No	Yes

(continued on next page)

| Table 4.10 | Quality *(continued)* |

Dependent Variable: Aspect (Dummy 1–0)							
			Quintiles	Urban		Rural	
Privatized (P)	0.1776***	–0.0979		–0.0927	0.0976*	–0.0234	0.0904
	–0.0524	–0.1600		0.0753	–0.0546	–0.1393	–0.0947
Quintile 1*P			–0.2714				
			–0.2392				
Quintile 2*P			–0.1403				
			–0.1997				
Quintile 3*P			–0.1808				
			–0.2126				
Quintile 4*P			–0.0513				
			–0.1599				
Quintile 5*P			–0.1045				
			–0.1871				
R^2	0.1197	0.1197	0.1239	0.1332	0.1332	0.0881	0.0881
No. Obs.	17569	17569	17569	16123	16123	1337	1337
TCI	No	Yes	Yes	No	Yes	No	Yes

Notes: Standard errors in parenthesis.
All regressions include the controls described in the text and in Table 4.7.
* significant at 10%, ** significant at 5%, *** significant at 1%

the fourth and fifth quintiles. Also, in rural privatized areas the frequency of the service is higher by a large percentage (i.e., around 20 percentage points higher; see Table 4.10, Panel B). The last quality of service variable is the subjective perception of the household with respect to the aspect of water.[12] In this case, for the whole sample water has better aspect in privatized areas, although the effect disappears when technical capacity is controlled for. However, the effect remains positive in urban privatized areas (Table 4.10, Panel C).

Another way to see the influence of municipal technical capacity is to estimate the effect of this variable in privatized municipalities (i.e., the interaction between technical capacity and the privatization dummy; see Table 4.11). These estimations show that the effects of privatizations in municipalities with better technical capacity are positive on access, prices and quality (measured as frequency of service). These results

---

[12] For this measure the ECV has data only for 2003.

Table 4.11	Privatization with Technical Capacity of the Municipalities		
	Access	Payment	Quality
	(Dummy 1–0)	(% of total income of the household)	Frequency of Service (Dummy 1–0)
Local Tax Revenues* Privatized	0.0006**	−0.0086***	−0.0213
	−0.0003	−0.0033	−0.0140
Saving Capacity* Privatized	0.0013***	−0.0171	−0.0844***
	−0.0004	−0.0191	−0.0145
Transfers* Privatized	−0.0001	0.0063	0.0415***
	−0.0003	−0.0112	−0.0072
R^2	0.58	0.04	0.32
No. Obs.	21003	16168	20296

Notes: Standard errors in parenthesis.
All regressions include the controls described in the text and in Table 4.7.
* significant at 10%, ** significant at 5%, *** significant at 1%

suggest that better technical capacity might result in better control (i.e., government regulation).

With respect to health, in general no effects of privatization on diarrhea are found.[13] In rural areas that privatized, however, the effect of privatization is a reduction of around 11 percentage points in the number of children with diarrhea. Although the effect is reversed when technical capacity is controlled for, the estimation loses degrees of freedom due to data availability[14] (Table 4.12, Panel A). Finally, with respect to the measure of weight for height, the effect of privatizations is also positive and significant when technical capacity is included as a control: municipalities that privatized have a higher measure of children's weight and height compared to the median than municipalities that did not privatize, and, in particular, urban areas. This result of variation over time is more

---

[13] Given that the DHS survey does not include questions about income, the estimations by quintiles are not included in this and the next table.

[14] Notice, however, that when technical capacity of the municipalities is included as control, the number of observations decreases significantly. This is due to the fact that, as stated above, the source of the dependent variable (DHS Survey) does include additional municipalities, with fewer inhabitants interviewed, and there is no data availability of measures of technical capacity for these new municipalities.

important since, as shown in the description of the variables, in privatized municipalities the measure of weight for height is lower than the median, and also in comparison to nonprivatized municipalities (Table 4.2, Panel B).

Conclusions and Policy Recommendations

In sum, Colombian water privatizations have had positive effects. Access has increased in privatized urban areas; water in privatized areas needs less treatment for use; the frequency of service in privatized areas is lower, but higher for the lower quintiles; the incidence of diarrhea in children in rural privatized areas is lower; and the measure of weight for height is

Table 4.12   Health

Dependent Variable: Children With Diarrhea in the Last 2 Weeks (%)						
			Urban		Rural	
Privatized (P)	−0.0470	−0.0640	0.1994	−0.0477	−0.1893***	−0.1098
	0.0503	0.0497	−0.1882	−0.0775	−0.0417	−0.2006
Time (t)	−0.0342***	−0.0469***	−0.0380***	−0.0497**	−0.0374**	0.0155
	0.0098	0.0219	−0.0121	−0.0229	−0.0167	−0.0809
P*t	0.0074	0.0043	−0.0051	0.0058	0.1191*	−0.1515***
	0.0184	0.0288	−0.0192	−0.0302	−0.0715	−0.0555
R^2	0.03	0.03	0.04	0.03	0.05	0.16
No. Obs.	9649	4958	6788	4551	2824	402
TCI	No	Yes	No	Yes	No	Yes
Dependent Variable: Weight for Height Standard Deviations from the Reference Median						
Privatized (P)	−0.2352	−0.4075	0.2129	0.2324	−0.6134**	−0.6606
	0.1732*	0.1355***	−0.6782	−0.2686	−0.2736	−0.6433
Time (t)	−0.0460	−0.0371	−0.0567	−0.0274	−0.0051	−0.4867**
	0.0284**	0.0627	−0.0358	−0.0458	−0.0659	−0.2121
P*t	0.0337	0.2693	−0.0006	0.2895*	0.2480***	0.2167
	0.0539	0.0819***	−0.0604	−0.1522	−0.0856	−0.3088
R^2	0.04	0.04	0.05	0.07	0.04	0.15
No. Obs.	9214	4735	6505	2709	4342	393
TCI	No	Yes	No	Yes	No	Yes

Notes: Standard errors in parenthesis.
All regressions include the controls described in the text and in Table 4.7.
* significant at 10%, ** significant at 5%, *** significant at 1%

higher in privatized municipalities. In addition, municipal technical capacity is important in generating these positive effects, potentially through better regulation. The negative effect related to payment seems to be more related to the elimination of cross-subsidies between the higher and the lower quintiles. Finally, the positive effects on health and quality in rural areas are outweighed by negative effects on prices and access.

The results of this chapter suggest several points for policy implementation. Since the effects of privatization are generally not homogeneous across income groups and across areas, the heterogeneity of impact has several implications. From the point of view of political economy, it is important to realize that the effects are not homogeneous across all households; it may therefore be useful to have in place targeted policies to mitigate negative effects. While this chapter finds overall positive outcomes from privatization, the unpopularity of privatization stems from other sources. These may include localized effects, mainly in the price system.

Privatization is apparently working better in urban than in rural areas. These effects are quite interesting, since it is believed that reaching the marginal house is more costly when the cost of provision of the service is already high. On the contrary, and despite higher levels of connectivity in urban areas, privatization is working in dense cities. The challenge is to expand the benefits of privatization to rural areas, and incentives to increase rural access may need to be generated through regulation.

It should be noted additionally that there are some cross-effects of privatization for low-income households. There are some positive effects on quality of service and, in general, on health outcomes that presumably have a greater impact on low-income families. In contrast, the price of service has been increasing more for low quintiles in privatized locations, and it appears likely that this price increase reflects the simultaneity of privatization and elimination of cross-subsidies. Despite the economic rationality of dismantling the subsidy systems, so that prices more accurately reflect the marginal cost of providing service, it may be preferable to improve the focus of cross-subsidies rather than eliminate them altogether. For example, Gómez-Lobo and Contreras (2003) discuss the differences between a means-tested system, such as Chile's, and

Colombia's geographical system, concluding that targeting by a means-tested instrument results in better focalization.

In terms of regulation, one of the main areas where regulation should focus is on quality measures of the service such as the frequency of service, water quality in terms of treatment, and the aspect of water. Second, regulation should generate incentives to increase access in rural areas.

CHAPTER FIVE

Stay Public or Go Private? A Comparative Analysis of Water Services between Quito and Guayaquil

Paul Carrillo, Orazio Bellettini and Elizabeth Coombs

M any Latin American countries face similar water problems: deteriorating systems and networks, lack of access to water and sewage for many of the populations' poorest, and governments without the resources or expertise to invest in change. Unfortunately, there is little consensus on how to solve these problems. Many countries, including Ecuador, have embarked upon various forms of privatization to increase investment in infrastructure and improve service provision and water quality. Most of the time, however, those privatizations have generated substantial controversy.

In the hopes of clarifying the true impact of water privatization—an issue in which ideology often overrides the facts—researchers have implemented numerous studies. Unfortunately, the results remain contradictory. On the one hand, the World Bank states that the number of water connections in Latin America has increased considerably since the onset of privatization, and that the venture as a whole has had "no major adverse impacts on poverty and inequality" (Leipziger, 2004). A number of authors have reported large gains in productivity and profitability associated with privatization in other sectors, providing evidence that points to privatization's achievements in increasing investment and productivity, as well as improving the access of poorer communities.[1] In

---

[1] In particular, see Megginson, Nash and van Randenborgh (1994); Barberis et al. (1996); Frydman et al. (1999); La Porta and López-de-Silanes (1999).

addition, privatization is shown to improve health indicators. For example, one study found that the expansion of water hook-ups as the result of water services privatization was associated with an 8 percent reduction in child mortality. Moreover, most of the reduction occurred in low-income areas (26 percent), where the network expansion was greatest (Galiani et al., 2005).

On the other hand, there have been many instances in both developed and developing countries where the privatization of water and sewage has faced serious challenges and has led to little, if any, net improvement in service provision. For example, structural adjustment reforms mandated by the International Monetary Fund (IMF) led to water price increases of as much as 200 percent in Cochabamba, Bolivia's third largest city, provoking widespread protests. In addition, international reviews in Europe and the United States do not find clear cost savings associated with privatization (Boyne, 1998; Renzetti and Dupont, 2003; Hodge, 2000), and privatization rates among municipalities in the United States have actually fallen since 1997, according to the International City/County Management Association (Warner and Hefetz, 2004). City officials cite problems with service quality and lack of cost savings, and statistical analysis of this reversal finds that problems with monitoring are key (Hefetz and Warner, 2004).[2]

Rather than finding new evidence against or in favor of privatization, this chapter documents the water provision experiences in the two largest cities in Ecuador: Guayaquil and Quito. Guayaquil, the largest city in Ecuador, has struggled to provide adequate water and sewage services to its residents. In the early 1990s, when water companies were run by the municipality, coverage rates were dramatically low. In 1994, the municipal government initiated a process of restructuring the public water and sewage companies and began to lay the groundwork for what would lead to the privatization of the sector. In 2001, the municipal water regulator

---

[2] For additional references that analyze the privatization of water services see Caplan, Evans and McMahon (2004); Clarke, Kosec and Wallsten (2004); Foster (2002); and Neu and Rahaman (2003). Other studies that analyze privatization processes more generally include Barberis et al. (1996); Bel and Warner (1996); Donahue (1989); Sclar (2000); and Shleifer (1998).

signed a 30-year integrated concession with Interagua, a subsidiary of the International Water Group, which handed over commercial risk and responsibility for the operations, maintenance, and administration of all potable water and sewage services in Guayaquil. On the other hand, in Quito, water-service providers have always operated under the direct authority of the municipality. Although they have historically performed better than the public companies in Guayaquil, the municipal water and sewage companies in Quito also faced financial management difficulties in the early 1990s. Instead of pursuing privatization, however, reforms were introduced to create more business-oriented practices.

To evaluate the effects of the privatization of water services in Ecuador, it is tempting to compare changes in water performance indicators (before and after privatization) between these two cities. In fact, several civic groups have criticized the terms of the concession process by making purely subjective comparisons between the performance of the private water company in Guayaquil and that of the publicly owned water company in Quito.[3] Such comparisons, however, could only identify the effects of privatization if Quito and Guayaquil were identical cities and if the privatization of water services were randomly assigned to the latter city. This is, of course, not the case.

This chapter has two main goals. The first is to provide an objective comparison of several indicators of water coverage, quality, and prices in both cities—both before and after the privatization of water services in Guayaquil. The type of data sources used make it possible to specifically control for income and, thus, to evaluate changes in water provision, particularly among the poor. These indicators provide useful information on how certain water-related features have changed over time and facilitate evaluating the performance of each company. The second goal is to document why such an exercise cannot identify the

---

[3] In the initial years of the concession there was very little negative public opinion about the process, but recently complaints about water quality have increased, and several civic groups have criticized the terms of the concession process. In some cases, arguments against the privatization of water services in Guayaquil have been made by comparing the performance of the private water company with the performance of a publicly owned water company in Quito. Most of the time, however, these comparisons have been merely subjective.

effects of the privatization of water provision. In particular, it is argued that before-concession water-coverage trends, rural-to-urban migration patterns, and other idiosyncratic institutional differences between these two companies may be driving a large portion of the quantitative results.

To assess how coverage levels have changed in Quito and Guayaquil, data are gathered from two national income and expenditure surveys that were administered in both cities in 1995 (before concession of water services) and in 2004 (after concession). A binary probit model and database are then used to identify the conditional probability that a household has access to water services. The estimates provide evidence that, in Guayaquil, households in the lowest income quintile have a 7 percent lower chance of receiving water services compared to 10 years ago (before the concession). In Quito, on the other hand, a household in the lowest income quintile in 2004 had a 3.5 percent higher probability of receiving water services than in 1994. These findings may suggest that in Guayaquil, water coverage among the poor has decreased during the past 10 years, both within the city and also relative to Quito.

To explore whether there have been changes in the quality of water services in these cities, the chapter further uses a detailed household employment survey that asks heads of households for their opinion about changes in water services (quality, pressure, and continuity) during the past six years. The survey was representative in both cities and included detailed information about household demographics, income, and employment. These data are analyzed using both an ordered probit and a linear model, and the results suggest that, on average, perceptions of how water quality and water continuity have changed during the past six years in Guayaquil do not differ statistically from perceptions in Quito. Interestingly, the poorest quintile of households in both cities consistently think that water quality has increased. This is not the case, however, with regard to water pressure. In all specifications, households systematically perceive that water pressure in Guayaquil has worsened relative to Quito.

Finally, the evolution of water prices is analyzed in Quito and Guayaquil during a 10-year period from January 1996 to July 2005, and it is found that the average price of water in Guayaquil has been higher after concession and has increased at a faster rate than in Quito, both in nominal and real terms.

The previous quantitative results are useful in evaluating the individual performance of each water company over time. As discussed above, however, they cannot identify the effects of privatization. In both cities, there may have been several other factors not necessarily related to the concession that affected water performance. These factors are discussed in the rest of the chapter.

First to be examined are before-concession water-coverage trends and rural-to-urban migration patterns. It is found that, from 1990 to 2000, Quito's coverage levels were increasing at a very high rate, while the opposite occurred in Guayaquil, particularly in its rural marginal areas. Furthermore, migration rates to Quito during the past six years have been higher than migration rates to Guayaquil for all consumption-level quintiles, with differences being higher for poorer families. Second, evidence is provided on the institutional differences between the companies in the two cities. Through interviews and focus groups, an institutional analysis of costs, quality, and performance is implemented that helps explain why Quito and Guayaquil have had such different experiences in water performance. The results suggest that Quito may not provide a suitable control group for identifying the effect of the privatization of water service in Guayaquil. While it is tempting to compare water indicators between Quito and Guayaquil, this exercise makes clear that the effects of water privatization on performance cannot be identified.

Background: Water Provision in Ecuador
Guayaquil

As the largest city in Ecuador, with a population of 2.5 million, Guayaquil has long struggled to provide adequate water and sewage services to its residents. In the early 1990s, its systems were in a state of near collapse. Financial mismanagement, inadequate maintenance and investment, and a history of overstaffing and political appointees all burdened its public utilities companies, which were heavily indebted and unable to provide basic services to the wave of unplanned communities that emerged with the construction of the *perimetral* (a highway that circles the city to alleviate vehicle congestion). Only 64 percent of the population had access to water service and only 46 percent had access to sewage in 2003 (Constance, 2003). With such low coverage rates, marginal communities were almost universally excluded from the official network and were left with no choice but to purchase water from *tanqueros* (private water-delivery trucks), which resulted in their paying 125 times more for water (sometimes up to 25 percent of their income) than those connected to the system (Ochoa and Prieto, 1995).

The First Steps toward Privatization

Recognizing a need for change, the municipal government initiated a restructuring process in 1994 that fundamentally changed the face of public utilities in Guayaquil. It merged the two previously separate public water and sewage companies under the auspices of the Empresa Cantonal de Agua Potable y Alcantarillado de Guayaquil (ECAPAG) and began to lay the groundwork for the privatization of the sector.

Upon the completion of the merger in 1996, ECAPAG set out to significantly reshape the provision of water and sewage services. By the late 1990s, it had streamlined staffing, created a division to respond to customer questions and complaints, and had improved efficiency in the operation of its services. The new ECAPAG began to expand distribution networks to marginalized areas of the city, rehabilitate treatment plants, purchase new, top-of-the-line equipment, and complete a process of extensive administrative and operational modernization.[4]

Although these public-system reforms were substantial, the level of indebtedness and the history of poor follow-through on loan obligations of the previous public utilities companies led international lenders to refuse access to credit for much-needed infrastructure construction without significant private sector involvement. While it was understood from the outset that ECAPAG would work toward implementing the long-term goal

---

[4] ECAPAG, undated document.

of privatization, it was not until 1995 that the ECAPAG directory officially approved the implementation process. In October 1997, Ecuador signed a loan contract with the Inter-American Development Bank (IDB) to finance the improvement of water and sewage services in Guayaquil—a contract that was contingent upon opening the sector to private concession. The loan granted by the IDB covered three principal areas: i) the concession process, ii) the transformation of ECAPAG into a regulatory body, and iii) the rehabilitation of the potable water and sewage systems.[5]

The Concession with Interagua

In 2001, ECAPAG signed a 30-year integrated concession with Interagua, a subsidiary of the International Water Group. This concession handed over commercial risk and responsibility for the operation, maintenance, and administration of all potable water and sewage services in Guayaquil, while maintaining ECAPAG as a regulator to ensure contract compliance. Key elements of the contract signed with Interagua include:[6]

- Operation and maintenance of potable water and sewage systems.
- Investment to improve quality of service.
- 5- and 10-year goals to improve minimum pressure, provision, and water quality.
- Investment to expand the system in the first five years.
- Obligation to install 55,238 new potable water connections and 55,238 new sewage connections in the marginalized sectors of the city.
- Obligation to invest $520 million in infrastructure, in addition to investment in rehabilitation and new connections, by the end of the 30-year contract.
- Obligation to install the number of new potable water and sewage connections necessary to reach 95 percent coverage of

[5] ECAPAG, undated document.

[6] List of contract obligations taken from ECAPAG, undated document.

> potable water and 90 percent coverage of sewage by the end of
> the second fifth-year period.
> - Beginning in the second fifth-year period, obligation to imple-
>   ment the new treatment plants and a macrosystem of drainage
>   as dictated by the Master Plan.
> - Obligation to respect and apply the pricing structure established
>   by ECAPAG for the first five years.
> - An investment goal of approximately $1 billion during the 30
>   years of contract.

The terms of the concession were designed specifically to avoid some of the pitfalls that arose under privatization schemes in other countries and, comparatively, can be considered relatively "poor friendly" in terms of coverage, price, and quality. Recognizing that one of Guayaquil's greatest problems has been the lack of inclusion of poor communities in the coverage of water and sewage networks, the concession contract specifically requires new connections to be provided to these communities at no cost. Emphasis in the first five years of the contract is placed on Interagua providing a minimum number of these new connections, and marginal communities are identified and incorporated according to an official expansion plan.

In order to avoid public backlash, as well as the burden of sudden price hikes, water tariffs are also strictly controlled for the initial years of the contract. The current tariff structure was designed and implemented by ECAPAG in the years prior to privatization both to more adequately cover the real costs of the system and to disassociate the changes in tariff structure from the privatization process so as to mitigate potential public backlash. For the first five years of the contract, the concessionaire is bound to uphold the pre-established pricing structure (barring any unforeseen changes on a national level that influence operational costs).

The concession contract also controls for water quality through the use of pressure and quality samples taken throughout the city and submitted to a number of laboratories for testing. Samples are taken by both Interagua and ECAPAG, and if they do not fall within the guidelines stipulated in the contract sanctions and fines can be imposed.

Public Response to the Concession

While there appeared to be very limited negative public opinion regarding the process in the initial years of the concession, the tide has recently turned, and the debate has become much more polarized. An increasing number of complaints have been emerging from marginal communities regarding poorer water quality and charges for services they do not receive. Despite the pressure and quality standards stipulated in the contract, residents of Guasmo Sur[7] consistently complain of turbid, foul-smelling water that is not fit for consumption, while residents of Suburbio Oeste[8] have struggled through both a hepatitis outbreak and periodic issues of decreased chlorine content/increased fecal content in the samples from their sector (Comisión de Control Cívico de la Corrupción 2005).

In early 2005, the Observatorio Ciudadano de Servicios Públicos, a citizens' watchdog group, was formed in an attempt to monitor the conduct and compliance of the concessionaire and regulator and to ensure a basic level of public and citizen accountability in the provision of the public good. Since its establishment, this group has issued numerous reports and conducted forums analyzing and criticizing both Interagua's compliance with the contract and ECAPAG's capacity as a regulator (Observatorio Ciudadano de Servicios Públicos, 2005a, 2005b, 2005c, 2005d, 2005e, and 2005f). In an attempt to better measure public opinion on issues related to water provision and public participation, the Observatorio organized a survey in November 2005 that polled more than 40,000 citizens, primarily from marginal sectors. Although the poll was voluntary and administered only in certain sectors of the city, the results overwhelmingly demonstrated a desire for systemic change and increased citizen participation.[9]

---

[7] Guasmo Sur is a low-income sector located in the southern part of the city.

[8] Suburbio Oeste is a low-income sector located in the western part of the city.

[9] According to this survey, more than 90 percent of respondents thought that Interagua and ECAPAG are not fulfilling their responsibilities to ensure quality water and sewerage services that are accessible to all citizens of Guayaquil. Furthermore, 95 percent believed

Critiques of the services rendered by Interagua have also come from the governmental watchdog entities the Defensoría del Pueblo and the Comisión de Control Cívico de la Corrupción (CCCC), both of which investigated the issue of water quality in Suburbio Oeste. In addition, from May to October 2005 alone, more than 400 articles related to water appeared in the Guayaquil press. In many cases, these articles made an implicit comparison between water services in Guayaquil and Quito to argue for or against the water services concession. Moreover, in most cases those comparisons were merely subjective. As noted above, such comparisons are not appropriate for identifying the effects of privatization, since several other factors not necessarily related to the concession may have affected water performance.

Quito

Water-service providers in Quito have always been under the direct authority of the public sector. EMAAP-Q, the municipal company that runs water and sewage services in Quito, was created in the mid-1990s by combining the former public water and sewage companies. EMAAP-Q encountered financial-management difficulties in the 1980s and early 1990s and was not able to meet citizen needs. However, reforms to create more business-oriented practices—such as cost savings and a stronger work ethic—increased efficiency and coverage. A detailed institutional analysis of Quito's water company is presented below.

Measuring Changes in Water Services between Quito and Guayaquil

An objective comparison of water services in Quito and Guayaquil requires the construction of an appropriate set of indicators. This section constructs

---

that the concession contract should be revised to contain clauses that guarantee the rights of all citizens of Guayaquil. Notice that, because of sample selection, these results are not representative of the whole population.

and analyzes the evolution of several indicators for water coverage, price, and quality in both cities over the 1995–2005 period.

Coverage

Data from two national income and expenditure surveys are used to explore household water coverage. Each survey consists of one representative (cross-section) sample of the urban population (about 12 cities) in Ecuador and was conducted by the Ecuadorian Institute of Statistics (INEC).[10] The first survey took place from August 1994 to August 1995, and the second in 2004. The surveys provide detailed information about household sources of income and expenditures for each respondent.[11] Descriptive statistics of this database are presented in Table 5.1.

The database compiled is a representative sample of the population of Quito and Guayaquil. The 1994 sample consists of 1,737 respondents (households) in Quito and 1,713 in Guayaquil, while the sample size in the 2004 survey is more than 40 percent higher in both cities.

Because this study is particularly concerned with identifying changes in water-coverage levels among those at the lower end of the income distribution, several variables were collected that provide information on household income level. In order to gather this information, households' real income was measured in terms of the number of representative baskets of goods and services (BGS) that they could buy with their total earned income.[12] According to the data shown in Table 5.1, mean household income (in real terms) decreased by about 4 percent in Quito and 20 percent in Guayaquil in the last decade.

---

[10] In Spanish, INEC stands for Instituto Nacional de Estadísticas y Censos.

[11] The INEC analyzed the structure of the households' expenditures to establish a representative basket of goods and services and to compute the Ecuadorian Consumer Price Index in both 1995 and 2004.

[12] The monetary value of the set of goods and services that a representative household spends money on to satisfy its basic needs is computed by the INEC on a monthly basis. The monetary cost of this set of goods and services was $362 in 1994 and $387 in 2004. INEC estimates are used to compute the real income of a household in the sample.

| Table 5.1 | Income and Expenditure Surveys by Region: Descriptive Statistics | | | | | | | |

| | Quito | | | | Guayaquil | | | |
	Mean	Std. Dev.	Min	Max	Mean	Std. Dev.	Min	Max
Year: 1994								
Water services	0.94	0.24	0	1.00	0.82	0.39	0	1.00
Income	1.94	2.77	0	38.88	1.94	2.67	0	31.75
Share of expenses in alimentation	0.24	0.19	0	0.99	0.29	0.20	0	1.00
No. of household members	4.21	1.78	1	15.00	4.84	2.03	1	16.00
No. of members below age 5	0.53	0.75	0	4.00	0.61	0.81	0	5.00
Years of education head household	10.12	4.96	0	22.00	9.01	4.89	0	22.00
Observations	1,737				1,713			
Year: 2004								
Water services	0.97	0.16	0	1.00	0.80	0.40	0	1.00
Income	1.87	1.31	0	10.94	1.55	1.08	0	7.30
Share of expenses in alimentation	0.32	0.18	0	0.84	0.45	0.22	0	0.91
No. of household members	3.78	1.71	1	14.00	4.36	2.21	1	26.00
No. of members below age 5	0.43	0.68	0	3.00	0.59	0.85	0	6.00
Years of education head household	10.51	4.89	0	20.00	9.32	4.65	0	20.00
Observations	2,460				2,819			

Other information was collected that describes the household's socioeconomic status, such as the share of expenses allocated to food, the number of people living in the same household, the number of children below the age of five, and the education level of the head of the household. Household size and number of children under the age of five are important variables because they are generally negatively correlated to income.

A binary probit model and survey data are used to identify the conditional probability that a household has access to water services in Quito and Guayaquil; the dependent variable equals one if the household is connected to the water network. The control variables include the household's income, the number of members in the family, and the head of household's education. Also included is an explanatory dummy variable that equals one if the survey was taken in 2004 (that is, the "privatization" dummy variable).[13]

To analyze the changes in water-service coverage before and after concession, the binary models are estimated with the sample data in Guayaquil and Quito in separate regressions. In addition, in order to identify changes in coverage level among the poor, the population is divided by per capita income quintiles, and the probit equation is estimated for each quintile. Results are presented in Tables 5.2 and 5.3, showing both the value of the coefficients and the marginal effects (evaluated at the sample mean of the independent variables in each income quintile).

The results for Guayaquil are shown in Table 5.2. In all income groups, the education level of the head of the household has statistically significant effects on the probability of having access to water services. This variable is most likely capturing the unobserved location of the housing unit, since higher-educated households tend to be located in neighborhoods that have better public services. The results also provide evidence that, with the exception of the richest quintile of the population, families with children under five years of age have had less chance of having access to water in Guayaquil after concession. This result is worrying considering that the health of young children is jeopardized by the lack of formal service.

The coefficients on the variable "privatization fixed effect" suggest that, on average, there are no significant changes in the probability of having access to water services before and after concession in Guayaquil. There is evidence, however, that households in the lowest income quintile have a lesser chance of receiving water services after concession. For example, the likelihood of these families obtaining water services decreased by approximately 7 percent in the past decade (Figure 5.1).

---

[13] Although the actual concession occurred in 2001, the 1994 survey is here referred to as providing "before concession" data, because it was the year before ECAPAG started undertaking drastic public sector reforms in preparation for the hand-over to a private company.

Table 5.2 Household's Per Capita Income Quintile, Guayaquil

	1		2		3		4		5	
	Coef.	Marginal Effect	Coef.	Marginal Effect	Coef.	Marginal Effect	Coef.	Marginal Effect	Coef.	Marginal Effect
Constant	0.047		−0.121		0.335*		−0.062		1.039***	
	(0.151)		(0.170)		(0.180)		(0.207)	(0.265)		
Privatization fixed effect	−0.188**	−0.066**	−0.071	−0.023	−0.079	−0.019	0.115	0.018	−0.335	−0.022*
	(0.083)	(0.029)	(0.090)	(0.029)	(0.106)	(0.026)	(0.136)	(0.022)	(0.212)	(0.012)
Number of household members	0.081***	0.029***	0.085***	0.027***	0.048	0.012*	0.176***	0.027***	0.032	0.002
	(0.021)	(0.007)	(0.025)	(0.008)	(0.030)	(0.007)	(0.049)	(0.007)	(0.059)	(0.004)
Number of household members < age 5	−0.18***	−0.064***	−0.151***	−0.049***	−0.229***	−0.057***	−0.233**	−0.036**	−0.089	−0.006
	(0.045)	(0.016)	(0.055)	(0.018)	(0.074)	(0.018)	(0.113)	(0.017)	(0.177)	(0.012)
Years of education head of household	0.038***	0.013***	0.066***	0.021***	0.065***	0.016***	0.074***	0.012***	0.071***	0.005***
	(0.012)	(0.004)	(0.012)	(0.004)	(0.012)	(0.003)	(0.013)	(0.002)	(0.017)	(0.001)
Number of observations	1,043		1,006		913		851		719	

Standard errors in parenthesis.
The covariance matrix was calculated using the White Heteroskedasticity–Consistent method.
* significant at the 10% level; ** significant at the 5% level; *** significant at the 1% level.

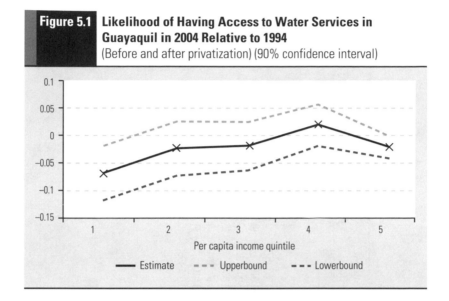

Figure 5.1 **Likelihood of Having Access to Water Services in Guayaquil in 2004 Relative to 1994**
(Before and after privatization) (90% confidence interval)

Coverage trends in Quito are quite different. Table 5.3 provides evidence that, in 2004, the probability of having access to water services in Quito has increased during the past decade. While these effects are particularly large and statistically significant for the third and fourth income quintile, even first income quintile households have notably increased the probability of being connected to the water network. For example, the results suggest that the likelihood of such low-income families obtaining water services increased by about 3.5 percent.

It is tempting to compare changes in water coverage indicators (before and after privatization) between these two cities and associate them with the effects of privatization.[14] However, such comparisons could only

---

[14] For example, one could use a "difference-in-difference" approach to explore the association between privatization and water coverage in these cities by estimating one binary model using data on both cities and adding "time," "city," and "private-owner" dummy variables to the set of other explanatory variables. A simpler and less rigorous comparison could consist of subtracting the marginal effects of the "privatization" dummy variable in Tables 5.3 and 5.4. For example, one may infer that, after privatization, a household in the first quintile in Guayaquil had a (0.035 – (–0.066)) = 0.10 lower probability of receiving water services than a similar household in Quito.

Table 5.3 Household's Per Capita Income Quintile, Quito

	1 Coef.	1 Marginal Effect	2 Coef.	2 Marginal Effect	3 Coef.	3 Marginal Effect	4 Coef.	4 Marginal Effect	5 Coef.	5 Marginal Effect
Constant	0.425642*		1.010865***		0.756489***		1.389733***		1.303794**	
	(0.229)	(0.268)	(0.251)	(0.372)	(0.555)					
Privatization fixed effect	0.233717	0.035377*	0.184751	0.017308	0.60336***	0.053657***	0.548409**	0.021615**	0.45211	0.004701
	(0.146)	(0.021)	(0.156)	(0.015)	(0.160)	(0.016)	(0.227)	(0.010)	(0.354)	(0.004)
Number of household members	0.045469	0.007189	−0.054566	−0.005046	0.031927	0.002408	0.005006	0.000153	−0.001535	−0.000012
	(0.036)	(0.006)	(0.042)	(0.004)	(0.049)	(0.004)	(0.059)	(0.002)	(0.124)	(0.001)
Number of household members < age 5	−0.099778	−0.015776	−0.021112	−0.001952	−0.266591**	−0.020104**	0.110608	0.003389	0.033424	0.000264
	(0.082)	(0.013)	(0.105)	(0.010)	(0.135)	(0.010)	(0.206)	(0.006)	(0.274)	(0.002)
Years of education head of household	0.100214***	0.015845***	0.10613***	0.009815***	0.072881***	0.005496***	0.040237	0.001233*	0.083316***	0.000658*
	(0.020)	(0.003)	(0.024)	(0.002)	(0.017)	(0.001)	(0.025)	(0.001)	(0.032)	(0.000)
Number of observations	704		740		834		893		1,026	

Standard errors in parenthesis.
The covariance matrix was calculated using the White Heteroskedasticity–Consistent method.
* significant at the 10% level; ** significant at the 5% level; *** significant at the 1% level.

identify the effects of privatization if Quito and Guayaquil were identical cities and if the privatization of water services were a random event. This is not the case. In both cities, there may have been several other factors not necessarily related to the concession that affected water coverage, such as previous water-coverage trends and rural-to-urban migration rates.

Water Quality

To measure whether the quality of water in Guayaquil has improved or worsened since the concession, the chemical and biological makeup of water samples taken before and after 2001 should be analyzed. Unfortunately, such data are not yet publicly available.

Information on water quality samples in Guayaquil has only been publicly available since October 2005, and these data were collected between that date and January 2006 to determine whether poor neighborhoods have lower water quality. The data include details on the chemical composition of the water samples and the address of the properties where the samples were taken. Individual addresses were used to match the water-samples database with poverty data from the Ecuadorian 2000 Census. While this analysis does not assess whether there have been any changes in water quality during the past 10 years, it does help to explain whether there are systematic differences in the quality of the water provided to the poor.

The ECAPAG web page included 291 water-quality test records; of those records, however, 10 files would not open, 37 files lacked necessary information, and 44 locations could not be matched with a specific address. A map of Guayaquil was then constructed that divided the city into census zones and quintiles of poverty level as determined by a national poverty index in the Census (NBI for its initials in Spanish). Finally, the water records were matched with the Census poverty levels. Not surprisingly, water quality was tested at the highest rate by far in the richest areas, and the number of sites tested fell steadily as the poverty level of the location increased.

The water-sample tests contained information on many variables (such as chlorine, turbidity, fecal residuals, bacterial analysis, and pH levels). The present focus is on chlorine and water-clarity indicators, because these are the only two available for most of the tested sites. Higher chlorine levels indicate larger amounts of disinfectant in the water, and lower turbidity indicates greater water clarity. As shown in Figure 5.2, wealthier areas generally had higher levels of chlorine and greater water clarity than poorer sectors. For example, poorer areas have an average of 20 percent less chlorine than other samples. However, those differences are not statistically significant.

A question of further interest is whether there have been changes in the quality of water services in Guayaquil relative to Quito after concession. For this purpose a detailed household survey was used, which contains information on individuals' perceptions of changes in various aspects of water quality during the past six years.[15]

The present study focuses on three measures of water services that are available in the Ecuadorian Monthly Employment Survey: overall water quality (purity, odor), water pressure, and water continuity. In the April 2006 survey, heads of households were asked: "In your housing unit, has the quality/pressure/continuity of water improved, remained constant, or decreased over the past six years?" The responses to these questions and a simple regression analysis are used to identify whether there are any systematic differences in these variables in Guayaquil relative to Quito.

Two types of regression models are used. The first is a traditional linear model and the second is an ordered probit model. The dependent variable is an integer that represents five different categories: 1) notably decreased, 2) decreased, 3) remained the same, 4) improved, and 5) notably improved. Also included are several independent variables that may explain the way individuals express their opinions about water issues. For example, it is possible that less-educated people are more optimistic about the future and perceive that water quality changes have been higher than in reality. For this reason, independent variables

---

[15] This survey is known as the Ecuadorian Employment Survey and is carried out by the Facultad Latinoamericana de Ciencias Sociales (Flacso)-Ecuador on a monthly basis in the three largest urban areas in the country. The survey is used by the Central Bank of Ecuador to compute employment statistics and consumer confidence indices.

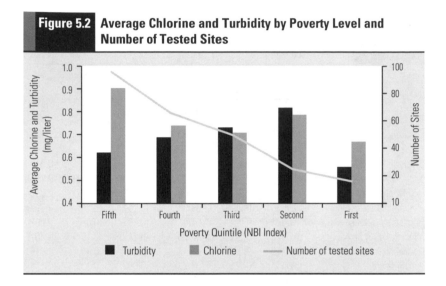

Figure 5.2 Average Chlorine and Turbidity by Poverty Level and Number of Tested Sites

are included such as the number of children (age five and younger) in the housing unit and head of household's age, gender, education, marital status, and employment status. Because effects among the poor are of particular interest, income is also controlled for by adding a set of dummy variables that describe the per-capita income quintile to which the household belongs.

For both the linear and the ordered probit models three different specifications are estimated. In the first and second specifications, only income and city effects are controlled for; the third specification adds several other independent variables to control for household demographics. The results are presented in Tables 5.4, 5.5, and 5.6.

The results suggest that, on average, perceptions of how water quality and water continuity have changed in Guayaquil during the past six years do not differ statistically from the same perceptions in Quito. However, households in the poorest quintile consistently think quality has increased. This is not the case, however, with perceptions of water pressure. In all specifications, there is evidence that households think that water pressure in Guayaquil has worsened relative to Quito. In particular, this effect seems to be stronger among those households in the second income quintile. These results are consistent with the fact that Interagua

Table 5.4	Changes in Water Quality during the Past Six Years

(Dependent variable: Has the quality of water improved, remained constant, or decreased during the past six years?)

	Linear Regression			Ordered Probit		
	(1)	(2)	(3)	(1)	(2)	(3)
Constant	2.838***	2.846***	2.959***			
	(0.039)	(0.046)	(0.145)			
Guayaquil fixed effect	0.046	0.021	0.021	0.091	0.04	0.041
	(0.035)	(0.077)	(0.079)	(0.070)	(0.151)	(0.155)
Household belongs to income quintile No 1	0.118**	0.072	0.068	0.233**	0.142	0.135
	(0.059)	(0.093)	(0.098)	(0.117)	(0.185)	(0.193)
Household belongs to income quintile No 2	0.082	0.043	0.04	0.161	0.085	0.082
	(0.057)	(0.079)	(0.086)	(0.113)	(0.154)	(0.167)
Household belongs to income quintile No 3	−0.012	−0.056	−0.054	−0.03	−0.12	−0.116
	(0.050)	(0.069)	(0.073)	(0.099)	(0.137)	(0.143)
Household belongs to income quintile No 4	−0.045	−0.005	−0.005	−0.092	−0.013	−0.014
	(0.052)	(0.068)	(0.070)	(0.103)	(0.135)	(0.138)
Guayaquil and household income quintile No 1		0.078	0.083		0.157	0.166
		(0.124)	(0.124)		(0.245)	(0.246)
Guayaquil and household income quintile No 2		0.082	0.096		0.161	0.189
		(0.116)	(0.119)		(0.229)	(0.234)
Guayaquil and household income quintile No 3		0.091	0.092		0.187	0.189
		(0.102)	(0.103)		(0.202)	(0.203)
Guayaquil and household income quintile No 4		−0.09	−0.089		−0.178	−0.176
		(0.106)	(0.107)		(0.209)	(0.210)
Household demographic variables[1]	No	No	Yes	No	No	Yes
Number of observations	1131	1131	1131	1131	1131	1131
R^2	0.01	0.02	0.02			

Standard errors in parenthesis.
The covariance matrix was calculated using the White Heteroskedasticity-Consistent method.
[1] Demographic variables include: the number of children (age 5 and younger) in housing unit, head of household's age, gender, education, marital status, and employment status.
* significant at the 10% level; ** significant at the 5% level; *** significant at the 1% level.

has made an effort to increase the number of water connections. Without increasing the production of water, water pressure inevitably worsens.

These results provide useful information on how households' perceptions of water quality in Quito and Guayaquil have changed over time. It should be emphasized, however, that these results do not identify any effect of privatization of water services.

Table 5.5	**Changes in Water Continuity during the Past Six Years** (Dependent variable: Has the amount of time when water services are available increased, remained constant, or decreased during the past six years?)

	Linear Regression			Ordered Probit		
	(1)	(2)	(3)	(1)	(2)	(3)
Constant	2.839***	2.852***	2.889***			
	(0.035)	(0.039)	(0.145)			
Guayaquil fixed effect	0.054	0.015	0.026	0.101	0.039	0.06
	(0.037)	(0.075)	(0.079)	(0.071)	(0.145)	(0.152)
Household belongs to	0.081	−0.076	−0.089	0.155	−0.139	−0.168
income quintile No 1	(0.063)	(0.097)	(0.098)	(0.121)	(0.187)	(0.190)
Household belongs to	0.114**	0.108	0.1	0.225**	0.221	0.206
income quintile No 2	(0.057)	(0.075)	(0.079)	(0.110)	(0.147)	(0.154)
Household belongs to	0.066	0.015	0.012	0.13	0.037	0.032
income quintile No 3	(0.051)	(0.064)	(0.069)	(0.100)	(0.127)	(0.135)
Household belongs to	−0.025	0.017	0.011	−0.049	0.038	0.027
income quintile No 4	(0.049)	(0.061)	(0.062)	(0.095)	(0.118)	(0.120)
Guayaquil and household		0.244*	0.245*		0.451*	0.456*
income quintile No 1		(0.132)	(0.133)		(0.253)	(0.255)
Guayaquil and household		0.025	0.025		0.034	0.036
income quintile No 2		(0.117)	(0.120)		(0.225)	(0.230)
Guayaquil and household		0.11	0.106		0.196	0.189
income quintile No 3		(0.106)	(0.109)		(0.206)	(0.211)
Guayaquil and household		−0.094	−0.098		−0.195	−0.205
income quintile No 4		(0.104)	(0.106)		(0.201)	(0.205)
Household demographic variables[1]	No	No	Yes	No	No	Yes
Number of observations	1131	1131	1131	1131	1131	1131
R^2	0.01	0.02	0.02			

Standard errors in parenthesis.
The covariance matrix was calculated using the White Heteroskedasticity-Consistent method.
[1] Demographic variables include: the number of children (age 5 and younger) in housing unit, head of household's age, gender, education, marital status, and employment status.
* significant at the 10% level; ** significant at the 5% level; *** significant at the 1% level.

Price

To analyze the price of water in Guayaquil before and after the concession, monthly average water prices in Guayaquil and Quito are compared over a 10-year period (January 1996 to July 2005). This highly reliable

Table 5.6	Changes in Water Pressure during the Past Six Years
	(Dependent variable: Has the pressure of water improved, remained constant, or decreased during the past six years?)

	Linear Regression			Ordered Probit		
	(1)	(2)	(3)	(1)	(2)	(3)
Constant	3.018***	3.044***	3.01***			
	(0.039)	(0.045)	(0.176)			
Guayaquil fixed effect	−0.1**	−0.177**	−0.166*	−0.171**	−0.305**	−0.287**
	(0.039)	(0.082)	(0.085)	(0.067)	(0.142)	(0.145)
Household belongs to	0.022	−0.105	−0.105	0.038	−0.182	−0.182
income quintile No 1	(0.069)	(0.095)	(0.099)	(0.117)	(0.164)	(0.170)
Household belongs to	−0.039	−0.135*	−0.134*	−0.068	−0.233*	−0.233*
income quintile No 2	(0.060)	(0.077)	(0.081)	(0.102)	(0.132)	(0.139)
Household belongs to	−0.074	−0.101	−0.1	−0.127	−0.175	−0.173
income quintile No 3	(0.056)	(0.073)	(0.076)	(0.096)	(0.125)	(0.131)
Household belongs to	−0.124**	−0.099	−0.097	−0.214**	−0.171	−0.168
income quintile No 4	(0.056)	(0.070)	(0.071)	(0.096)	(0.120)	(0.121)
Guayaquil and household		0.221	0.21		0.382	0.363
income quintile No 1		(0.138)	(0.140)		(0.237)	(0.241)
Guayaquil and household		0.208*	0.208*		0.357*	0.358*
income quintile No 2		(0.123)	(0.125)		(0.211)	(0.214)
Guayaquil and household		0.08	0.069		0.138	0.12
income quintile No 3		(0.116)	(0.117)	(0.199)	(0.201)	
Guayaquil and household		−0.044	−0.056	−0.075	−0.096	
income quintile No 4		(0.117)	(0.117)	(0.200)	(0.200)	
Household demographic variables[1]	No	No	Yes	No	No	Yes
Number of observations	1131	1131	1131	1131	1131	1131
R^2	0.01	0.02	0.02			

Standard errors in parenthesis.
The covariance matrix was calculated using the White Heteroskedasticity-Consistent method.
[1] Demographic variables include: the number of children (age 5 and younger) in housing unit, head of household's age, gender, education, marital status, and employment status.
* significant at the 10% level; ** significant at the 5% level; *** significant at the 1% level

information comes from monthly surveys undertaken by INEC with the objective of calculating the Consumer Price Index.[16]

---

[16] Pricing information taken from the water companies may be biased, since they may have incentives to provide misleading information. On the other hand, information collected by the INEC should be reliable.

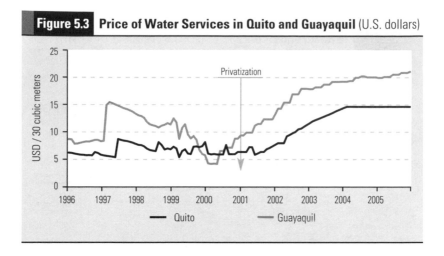

Figure 5.3 **Price of Water Services in Quito and Guayaquil** (U.S. dollars)

These surveys are used to analyze the evolution of the average cost of 30 cubic meters of water in Quito and Guayaquil. Figure 5.3 presents the amounts in current U.S. dollars. From the beginning of 1996 until October 1999, water was on average 66 percent more expensive in Guayaquil than in Quito. On the other hand, from November 1999 to May 2000, water was 25 percent more expensive in Quito than in Guayaquil. This tendency reversed a few months after the concession, however, and the price of water in Guayaquil again surpassed the price of water in Quito. Thus, there exists evidence that after the concession, the average price of water in Guayaquil has increased in comparison to Quito. The price gap jumped to $3 (per-unit) immediately after concession and climbed steadily to more than $4 by 2005.

To calculate the evolution of the price of water in real terms, the nominal price of water in Quito and Guayaquil is divided by the Price Index in each city. The index is then standardized so that the price of water in Quito in January 1996 is equal to 100. The corrected evolution of prices is represented in Figure 5.4. As in the preceding figure, the difference between the relative price of water in Guayaquil and the price in Quito is positive at the beginning of 1996 and decreases until a few months before privatization. Approximately six months before the concession, the difference becomes positive again and remains higher than Quito for the following years.

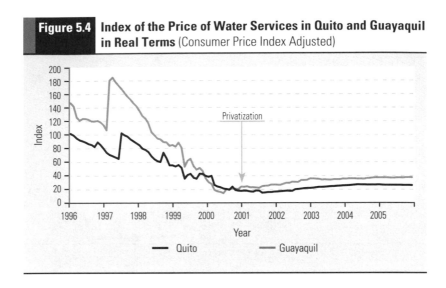

Figure 5.4 **Index of the Price of Water Services in Quito and Guayaquil in Real Terms** (Consumer Price Index Adjusted)

It should be acknowledged that changes in nominal water prices in Guayaquil have been limited so far by the concession contract. Once the first five years of the concession are over, Interagua will be able to change tariffs without ECAPAG approval.

Identifying the Effects of the Privatization of Water Services

If Quito and Guayaquil were identical cities and the privatization of water services were randomly assigned to the latter city, some of the estimates in the previous section could be used to measure the effects of privatization on water coverage, quality, and prices. This is, of course, not the case. Quito and Guayaquil are very different cities, especially in terms of water provision.

This section first describes how differences in previous coverage trends and migration patterns between Quito and Guayaquil can bias the quantitative results. A detailed comparative institutional analysis of both water companies is then provided, which leads to the conclusion that they face radically different environments and that Quito is not a suitable control group for identifying the effects of water concession in Guayaquil.

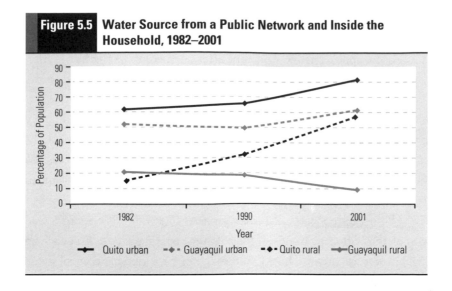

Figure 5.5 **Water Source from a Public Network and Inside the Household, 1982–2001**

Previous Trends in Water Coverage

Figure 5.5 presents data on water-coverage levels in Quito and Guayaquil from three National Population and Housing Censuses. Unfortunately, the Census data do not allow the analysis of water-coverage levels by income quintile; it is only possible to determine whether residences are located in rural or urban areas within the borders of these cities. However, because rural areas in both cities are primarily populated by poor families, it is reasonable to assume that coverage levels in rural areas are somewhat similar to the coverage levels of families that belong to the first income quintile.

Using a simple linear extrapolation of the Census data, it is estimated that by 2004 (the year of the income and expenditure survey), water-coverage rates in the rural areas of Guayaquil would have been 6.1 percent lower than corresponding coverage levels in 1994.[17] Since

[17] According to the Census, water-coverage levels in rural Guayaquil were 18 percent in 1990 and 9.8 percent in 2001, that is, they decreased at an average annual rate of 5.68 percent. Using this average annual decrease rate, it is estimated that the corresponding water-coverage rates in 1994 and 2004 may have been 14.2 percent and 8.3 percent, respectively. The difference between the two is 6.1 percent.

the previous section provided evidence that poor households (first in-come quintile) in Guayaquil decreased their likelihood of obtaining water services by approximately 7 percent after concession, previous trends in water-coverage levels may explain a large portion of the decreased access to water among the poor.

Coverage levels in Quito's rural areas, on the other hand, show an opposite trend. Quito made significant improvements, particularly, during the 1990s. In 1990, water-coverage levels reached 32 percent; by 2001, they increased to 56 percent, and by 2004 they may have climbed to 65 percent.[18]

Before the concession of water services, Quito and Guayaquil showed completely different trends in their water-coverage levels, especially in the cities' rural areas. Thus, Quito may not be a suitable control group for identifying the effects of privatization of water services, particularly among those at the low end of the income distribution.

Migration Trends and Urban Planning

To understand changes in coverage levels between Quito and Guayaquil, it is also important to analyze the two cities' migration trends. For example, higher rates of migration to Guayaquil may explain why water coverage decreased relative to Quito.

Data from the 2004 Demographic Survey on Mother and Infant Health (ENDEMAIN)[19] were used to identify migration rates. In this

---

[18] Using a similar linear extrapolation, it may be inferred that water-coverage levels in rural Quito increased from 39 percent in 1994 to 65 percent in 2004.

[19] Since 1987, the Center for Population Studies and Social Development in Ecuador (CEPAR, for its initials in Spanish) has periodically published its Demographic Survey on Mother and Infant Health, the ENDEMAIN. The last survey was taken in 2004. Through the compilation of numerous statistics related to health care, infant and child mortality, reproductive health, fertility, domestic violence, sexually transmitted diseases, household demography, migration, and other topics, CEPAR attempts not only to describe, but also to identify any patterns or problems in terms of access to and usage of health services. Socioeconomic status (based on consumption), for example, is one of the household statistics collected to ascertain if persons of lower socioeconomic status have less access to health care or have to spend more to obtain it than persons of higher socioeconomic status. The information is collected through household and individual surveys, with responses

Table 5.7	**Migration Patterns in Quito and Guayaquil**

(Answer to the question: Where did you live in January 1999?)

	Consumption Quintile					
	1	**2**	**3**	**4**	**5**	**Overall**
Quito						
Number of Respondents	290	389	416	580	580	2,255
% Lived in same city	78%	86%	86%	90%	95%	88%
% Lived somewhere else	22%	14%	14%	10%	5%	12%
Guayaquil						
Number of Respondents	203	433	541	500	375	2,052
% Lived in same city	92%	92%	98%	97%	93%	95%
% Lived somewhere else	8%	8%	2%	3%	7%	5%
Difference in migration rates between	0.145	0.058	0.121	0.074	-0.013	0.070
Quito and Guayaquil	(0.031)	(0.022)	(0.018)	(0.014)	(0.016)	(0.008)

Source: ENDEMAIN 2004.
Standard errors in parenthesis.

survey, heads of households in Quito and Guayaquil were asked if they had resided in the same city in 1999.[20] Furthermore, the survey contains information on households' consumption and, therefore, we were able to rank households by their per-capita consumption level. This information is displayed in Table 5.7.

Clearly, migration rates to Quito are higher than migration rates to Guayaquil for all consumption-level quintiles, and these differences are higher for poorer families. Whereas only 8 percent of the poorest families in Guayaquil lived in another city five years earlier, 22 percent of the poorest families in Quito had recently migrated. Table 5.7 thus provides evidence that decrease in coverage in Guayaquil after concession relative to Quito could not be explained by migration trends to these

---

solicited from 17 different sections of the country. Of the 10,966 observations in the 2004 ENDEMAIN, 1,151 come from Quito and 957 come from Guayaquil. For more details on the ENDEMAIN, see Angeles, Trujillo and Lastra (2005).

[20] Although it would have been optimal to observe migration trends during the past 10 years in order to make them comparable with previous results, such information was not available.

cities, providing further additional evidence that Quito does not provide a suitable control group.

Other Institutional Factors

This section analyzes the institutional environments of both companies. To understand the institutional contexts of these entities, interviews were conducted with employees of EMAAP-Q, ECAPAG, Interagua, the University of San Francisco at Quito, the Inter-American Development Bank and the Observatorio Ciudadano de Servicios Públicos. In addition, focus groups were conducted with residents of various marginal communities throughout Guayaquil to obtain first-hand accounts of their experiences with water services.[21]

External Factors that Contribute to Performance

Geographic characteristics. Both cities have geographic advantages and disadvantages in terms of capturing, cleaning, and distributing water. Quito, for example, has a more complicated system of carrying water from its sources to the city and its suburbs. Hundreds of kilometers of piping are used to transport water from seven water-capture sites to 22 treatment plants. Guayaquil, on the other hand, has only one source, the River Daule, and only three treatment plants, which are relatively close in proximity. One could argue that Quito's many capture and treatment sites complicate coordination and raise the cost of water production.

However, the mountains surrounding Quito provide it with an advantage in terms of water pressure. The majority of the 172 water tanks are located above the valley where Quito is located, and EMAAP-Q consequently has to restrain rather than create water pressure. In contrast, Guayaquil's relatively flat topography, combined with the fact that its treatment plants are located above the river, means that money has to

---

[21] Details about the interviews and focus groups can be found in Appendix 1 of Carrillo, Bellettini and Coombs (2007), the working paper on which this chapter is based.

be invested not only in pumping water to the plant, but also in distributing it throughout the system.

Probably the most important geographic characteristic that contributes to water entity performance is the quality of the water *before* it arrives at the treatment plants. Quito is proud that its water tests well below the water-turbidity requirement of 5 units, reaching the treatment plants at around 1 or 2 units. The water consumed by Quiteños comes from melted snow and glaciers from the various volcanoes surrounding the municipality—a great advantage not shared by Guayaquileños. In fact, Guayaquil's treatment plants have to undertake extensive sedimentation and filtration processes in order to reduce the organic material and contamination caused by nearby factories, boats, and communities. Unfortunately, heavily treated water does not always guarantee lasting quality because the effects of the chlorine used in the treatment dissipate over time and as the water moves through the water mains.

Water Companies and Political Influence

According to several interviewees, Guayaquil's political environment has been one of the primary causes of the poor historical performance of its public water providers. According to those individuals, past (pre-privatization) water-service providers in Guayaquil did not operate as independent and technical entities, but as mechanisms for mobilizing political power, thus damaging their institutional capacity and the provision of services.

It was not until the establishment of ECAPAG in 1995 that overstaffing and political influence began to come under control. The national law establishing ECAPAG removed it from the influence of local politics and gave it the autonomy to carry out the provision of water and sewage services in an apolitical, technical manner. In preparation for the privatization process, ECAPAG began to streamline staffing as well as implement numerous administrative and organizational reforms. To this day, the municipality plays a very limited official role in water provision, and the mayor has only a single representative on the ECAPAG board.

In contrast, interviewees in Quito emphasized the technical—not political—nature of EMAAP-Q. Although low turnover rates are positive,

Quito may also be suffering from overstaffing (see Table 5.8), indicating that it is not as efficient and apolitical as it may seem at first glance. Although EMAAP-Q has not been heavily politicized, the mayor plays an active role in the company's decisions.

Measures of Efficiency and Current Performance

Several indicators are chosen to measure the current institutional efficiency and performance of both water companies: 1) response to consumer complaints, 2) payment rate, 3) ratio of employees per 1,000 connections, 4) percentage of water lost, 5) general finances and 6) management and technical innovations. These indicators may provide a better sense of how efficiently and effectively the public and private institutions are working.

Response to consumer complaints. Water-providing entities in both Quito and Guayaquil have mechanisms for receiving, processing, and addressing consumer complaints, but Quito's appear to be more responsive and user-friendly. In Quito, there are two main systems by which to make a complaint—via telephone or via an EMAAP-Q office. There are 12 client services centers throughout the municipality, where 83 percent of the complaints are made. These offices are only open on business days from 7 a.m. to 7 p.m., but a call center additionally takes complaints 24 hours a

Table 5.8 Management Indicators from Guayaquil's and Quito's Water Companies

Indicator	Guayaquil Before Concession	Present	Quito 1996	Present
Number of employees/ 1000 connections	9.4	3	7.1	6.1
Percent of water lost due to leaks in the system or nonpayment	79%	68%	Not Available	30%
Payment rate	50% (1996) 76% (2001)	84%	62%	79%
Percent connections with meters	24%	49%	67%	97%

day. Both methods are connected to the AS400 system, which processes and distributes the complaints to the appropriate areas. According to the director of client services, there has never been a complaint related to the quality of water. Most have to do with billing problems, which are addressed by a home visit on the next working day. New connections are addressed within 20 days and reconnections within 48 hours after payment. Pipe breaks and technical complaints have response rates of 24 hours within the city of Quito and 48 hours in the parishes. In 2002, EMAAP-Q set a goal to keep the annual complaint rate below 1 percent of the number of consumers, a goal it surpassed in 2005 with a complaint rate of 0.35 percent.

In Guayaquil, both Interagua and ECAPAG have established fairly thorough mechanisms for accepting and processing complaints, and the absolute number of complaints has actually gone down since the concession. Under ECAPAG, prior to privatization, the total number of complaints received per month averaged approximately 2,000, while Interagua now averages roughly 800. Complaints related specifically to billing have also decreased. In 2001–2002 there were approximately 0.6 complaints related to billing per account, a number that has now decreased to 0.4.

Processes for receiving and attending to customer complaints have also expanded since the concession. Similar to EMAAP-Q, Interagua accepts complaints either through its call center, which is open from 7 a.m. to 8 p.m. Monday to Friday and with limited hours on Saturday, or through its two customer service centers in the northern and southern parts of the city, which are open from 8:30 a.m. to 4 p.m. Unfortunately, neither the Interagua nor ECAPAG call centers offers a toll-free number.

Interagua responds to first-instance billing complaints within 30 days and second-instance complaints within 10 days. Technical complaints have a much faster response time, which varies according to their gravity. Reconnections are made within 48 hours of payment.

As the operator and provider of water services, Interagua has the primary responsibility for addressing consumer complaints and concerns. However, as the regulating entity, ECAPAG monitors the database of complaints received and responded to by Interagua and also acts as a third-instance appeal. When a consumer makes a complaint, the first two instances go through Interagua. However, if the consumer remains dissatisfied with the decision, he or she may file an appeal with ECAPAG, which then conducts its own investigation, reviews the prior decisions, and makes its own decision. Like Interagua, ECAPAG has initiated a call center to receive customer complaints and concerns and responds to such complaints within 30 days.

While mechanisms for accepting and responding to complaints exist, focus groups conducted in marginal areas of the city reported high levels of frustration with Interagua's responses to user complaints and requests.[22] Participants reported poor treatment by Interagua employees working in both the customer service centers and in the field. Others felt that they were "given the run-around" when they attempted to get a response to the complaints they had filed.

Payment rates. In Quito, the payment rate has increased substantially in the past 10 years, from 42 percent in 1996 to 80 percent in 2004 (Table 5.8). Although the 20 percent still outstanding is high, it represents what is owed by only 3 percent of total customers.

In Guayaquil, payment rates have also increased substantially—both since ECAPAG assumed control of water provision operations and in the four years of Interagua's control. In the years prior to the creation of ECA-PAG, payment rates averaged only 50 percent, not enough to cover even the basic costs of water provision. Just before the concession, ECAPAG had reached a payment rate of approximately 76 percent. According to one of ECAPAG's directors, since the concession, payment rates have continued to rise to 84 percent, with additional increases each year. In addition to accepting payments in Interagua offices, the company has worked to expand the methods through which monthly bills can be paid.[23]

Ratio of employees per 1,000 connections. An indicator commonly used to determine efficiency and management performance is the number

---

[22] Specific dates and locations of the focus groups are presented in Appendix 1 of Carrillo, Bellettini and Coombs (2007).

[23] C. Espinoza, *Director de Control y Regulación Económica*, ECAPAG. Interviews conducted November 17, 2005 and March 17, 2006.

of employees per 1,000 connections.[24] As shown in Table 5.8, Quito has approximately 6 employees per 1,000 connections, down from 7.1 in 1996. Interagua, on the other hand, meets an industry standard of efficiency with a ratio of 3 employees per 1,000 connection. In this area, privatization has definitely had an impact on efficiency, considering that the pre-privatization water providers in Guayaquil had extremely high employee-per-connection ratios.[25]

Physical and commercial losses. Another common measure of efficiency is the level of physical and commercial losses suffered by the water-provision companies, such as the amount of water unaccounted for either due to illegal connections, breaks in the system, or nonpayment. In 2005, EMAAP-Q lost about 30 percent of its water (Table 5.8).

Heavy losses in both of these areas have been one of the most serious problems in Guayaquil's water provision system. Under the provincial water provision entity EPAP-G, which existed prior to ECAPAG, water losses reached 75 percent—one of the highest rates in Latin America (IDB, 1996). Since Interagua has assumed operations and begun to invest more heavily in system infrastructure and new connections, the level of loss, although still high, has begun to decline. It is now at approximately 67 percent. Losses also originate from clandestine connections to the official water network. At this time, there is no truly accurate estimate of the number of clandestine connections in Guayaquil. However, many have been created by low-income residents who have water mains near their house but who have never been provided with formal connections to the system. Interagua currently has a program to regularize such users once their connections are discovered, and as Interagua expands the number of new connections to the more marginal sectors of the city the

---

[24] Based on technical criteria, experts typically recommend a ratio of approximately 2 to 3 employees per 1,000 connections in this type of industry.

[25] In 1982, EMAP-G (one of the previous municipal water providers) had a ratio of 15 to 1,000 and at the end of 1994, before control was turned over to ECAPAG, EPAP-G (the provincial water company that followed EMAP-G) had a ratio of 9.4 to 1,000 (Swyngedouw, 2004).

number of clandestine connections and concomitant losses is expected to decrease.

General finances. As of 2005, EMAAP-Q has an annual budget of $200 million. The budget has grown by approximately $80 million in the past five years. Expenditures are divided as follows: 60 percent project investment, 10 percent debt repayment, 20 percent operations and maintenance, and 10 percent administrative costs. EMAAP-Q earns its income from various sources: 60 percent service sales, 25 percent telecommunications taxes, 10 percent international loans, and 5 percent sales of electricity. The last item is perhaps the most interesting and innovative. Due to the fact that EMAAP-Q works with rushing water sources, it developed a system to generate electricity, earning almost $5 million in 2004. Originally, the electricity created was used within EMAAP-Q. However, with increased production it is now sold for additional income.

In terms of loan history, EMAAP-Q has worked with several international creditors for more than 15 years. EMAAP-Q is proud to have a direct loan relationship with these creditors, which is not common in Ecuadorian public agencies, most of which must work through the central government.

Unfortunately, information on Interagua's general finances was not publicly available. According to ECAPAG, Interagua's annual budget is $80 million, with approximately $40 million going to general operating expenses and $25 million to investments. How Interagua allocates funds within these general categories and how it makes use of the remaining $15 million is unexplained.

One possible explanation for performance differences between EMAAP-Q and Interagua may be related to their budgets. In the past 10 years EMAAP-Q has had a substantially higher budget than that of ECAPAG and Interagua. In particular, the telecommunications tax is said to give EMAAP-Q extra resources that allow it to afford higher employment levels.

Management innovations. In addition to selling electricity, other important technological innovations have been undertaken. One was the implementation of a digital mapping system in Quito in 1998. Now that 100 percent of the commercial area of Quito has been mapped, technicians can more easily identify breaks or problems in the system. With these technological improvements, technicians in charge of water production can measure the amount of water being produced at each capture site to divert it more effectively to processing plants. Another innovation has involved equipping treatment plants with a system to automatically and constantly test and treat the water.

In regard to administrative innovations, EMAAP-Q had to develop an effective communications system to allow all 22 water plants to coordinate activities. Furthermore, EMAAP-Q uses its long-term strategic plan to calculate tariffs. Instead of basing the price of water on current costs, EMAAP-Q includes costs of future projects. Additionally, EMAAP-Q is aware of the advantages of using private companies to implement some activities, such as meter reading and the printing of bills. Finally, 97 percent of customers now have meters, allowing EMAAP-Q to charge consumers more fairly. The meters are changed every five years to avoid mistakes.

While Interagua has implemented numerous reforms and innovations since assuming control of operations in 2001, the modernization of Guayaquil's water services first began under the direction of ECAPAG in the late 1990s in preparation for opening the sector to private concession. In the years leading up to the concession, ECAPAG expanded the network, incorporating 43,000 new users from marginal areas in Guayaquil, conducting numerous projects to rehabilitate and optimize the functioning of the water and sewage system, and designing a plan for providing potable water to Isla Trinitaria, a marginal suburb of Guayaquil. ECAPAG also implemented numerous reforms to modernize its administrative and operational systems, including automation of the billing system and significant improvements in customer service.

As a subsidiary of a multinational company with access to the latest technological developments and innovations, Interagua has continued to build upon the modernization projects first begun under ECAPAG. However, because the relevant information is not available to the public, the number of innovations implemented by Interagua since 2001 is not quantified here.

Other Factors Contributing to Institutional Strength

Civil society participation. There are various indicators that can be used to measure civil society participation in water provision: 1) newspaper articles critiquing or exposing water company performance, 2) civil society movements or organizations that disseminate information to citizens and press for reforms, 3) mechanisms provided by the water companies to receive feedback from citizens, and 4) programs promoted by the water companies to involve citizens in project activities.

In Quito, general participation is low in the first three areas. In terms of citizen participation in program activities, the Department for Public Works for Social Development of EMAAP-Q provides the materials and technical support to make free connections to water and sewage systems if the community provides the manual labor.

During 2005, the issue of citizen participation in the provision of Guayaquil's water services began to gain attention. In part due to a number of scandals regarding water quality, as well as media attention generated by a newly formed watchdog group, the number of newspaper articles focusing on water company performance skyrocketed. From just May to October 2005, more than 400 water-related articles were printed in the local press.[26] In addition to this increased media attention, and perhaps a cause of some of it, a citizen watchdog group (the Observatorio Ciudadano de Servicios Públicos) was formed to publicly monitor Interagua and ECAPAG and hold them accountable for the services they provide.

The Observatorio says it does not necessarily oppose the privatization of public services, but believes that adequate citizen oversight and accountability need to accompany such processes. Although the Observatorio is a relatively new organization, and nearly unique in Ecuador, it began operations by publishing numerous reports analyzing and critiquing the terms of the concession contract, the structure of ECAPAG, and the master plan for water and sewage development that was recently presented by Interagua.

---

[26] Articles compiled by the Observatorio Ciudadano de Servicios Públicos.

While the Observatorio's efforts have definitely created a public dialogue around the quality of services being provided by Interagua and the responsibilities that ECAPAG holds as a regulator, official channels for citizen participation within both Interagua and ECAPAG remain severely limited.

Regulatory Bodies. In Ecuador, water provision has virtually no regulation at the national level.[27] There exist, however, several forms of self-regulation at the municipality level. For example, various members of EMAAP-Q mentioned that the entity is "self-regulated." EMAAP-Q tests the quality of water in a laboratory that works independently from the Department of Water Production. However, it is located on the same site, if not in the same building, and the laboratory receives almost 100 percent of its funding from EMAAP-Q. Nonetheless, according to the director, the independence of his laboratory is respected.

Guayaquil, on the other hand, has a much more elaborate and established regulatory system, as is essential under privatization. Specific areas of regulation include the following:

a) *Client Attention*—ECAPAG monitors billing, consumer complaints, and Interagua's responses to those complaints. Monthly reports as well as Interagua's databases are submitted to ECAPAG for review. Goals in this area focus on follow-through and response time in addressing complaints, not on the total number of complaints made.

b) *Water Quality*—ECAPAG conducts monthly/weekly regulation of water quality throughout the city by subcontracting a number of laboratories to do four counter samples in various sectors. In addition, ECAPAG is immediately informed of any complaint made to Interagua regarding water quality and sends its own team out to take counter samples.

---

[27] For example, a new code of water-quality norms, which raised national standards, was passed in October 2005. However, no national body exists to hold municipalities accountable.

c) *Investments*—ECAPAG specifically regulates Interagua's completion of the required number of water/sewage connections stipulated in the concession contract. Because additional infrastructure investments are not required to go through any specific regulation or approval process, Interagua's follow-through in this area is indirectly regulated through its ability to meet service goals (such as pressure and continuity).

d) *Finances*—ECAPAG conducts regular reviews of Interagua's financial statements to ensure that it is in good financial standing for the projects it seeks to undertake. The contract does not stipulate any further regulatory influence over company finances.

Conclusions

The comparative study of Quito and Guayaquil's water-provision systems over a recent 10-year period provides several important contributions. The first, a battery of quantitative indicators of water performance in these two cities, offers evidence that water-coverage levels in Guayaquil have decreased during this period, particularly among those at the lower end of the income distribution. The opposite is true in Quito. Furthermore, households systematically perceive that water pressure in Guayaquil has worsened relative to Quito. Finally, in both nominal and real terms the average price of water in Guayaquil is higher and has increased at a faster rate than in Quito. Although these indicators provide useful information to evaluate the performance of the water companies over time, they cannot necessarily be used to assess the effects of privatization per se.

The chapter's second contribution involves documenting several differences between these two companies that should be considered when interpreting any of the quantitative findings. First, before-concession water-coverage trends and rural-to-urban migration patterns are radically different in both cities. Thus, relative changes in coverage levels between Quito and Guayaquil cannot be attributed to privatization. In addition, an institutional analysis of the two entities to measure both the external and internal management factors shows that these two entities face significantly different environments, and institutional differences alone may explain a large portion of the results obtained in the statistical analysis.

While it is tempting to compare water indicators between Quito and Guayaquil, it is clear that the effects of water privatization on performance cannot be identified. Further research is needed to evaluate these effects.

CHAPTER SIX

Access to Telephone Services and Household Income in Poor Rural Areas Using a Quasi-Natural Experiment in Peru

Alberto Chong, Virgilio Galdo and Máximo Torero

rivatization was supposed to deliver the goods. Some researchers claim that, indeed, it has. Recent evidence shows that firms have dramatically improved performance following privatization, and that such positive changes are the result of significant restructuring efforts. The empirical record shows that privatization leads to increased profitability and productivity, firm restructuring, output growth and even quality improvements (Chong and López-de-Silanes, 2004).[1] However, several critics claim that privatization has impacted consumer income and welfare negatively through decreased access, poorer distribution, and lower quality of goods and services (Bayliss, 2001; Birdsall and Nellis, 2002). These concerns are significant because, for the most part, the poorest segments of society are the main consumers of goods and services previously produced by state-owned enterprises.

Especially in the case of services and public utilities, access and distribution may be a concern, as some segments of the population may lack a way of entry to networks and thus may be unable to purchase these services independently of their price. The quality of services such as water, electricity, telecommunications or transportation may be reduced to try to

---

[1] Most cases of privatization failure may be linked to poor contract design, opaque processes with heavy state involvement, lack of re-regulation, and a poor corporate governance framework. In fact, it appears that firms undergo harsh restructuring processes following privatization and do not simply mark-up prices or lower wages (Chong and López-de-Silanes, 2004).

meet price regulation, for example. In all of these circumstances, consumer welfare may suffer as a result of privatization. For instance, Bayliss (2002) points to examples of botched privatizations in Puerto Rico and Trinidad and Tobago, where water privatization led to price hikes and no apparent improvement in provision. Similarly, the privatization of the electric sector in the Dominican Republic is claimed to have led to more blackouts and higher utility prices, culminating in civil unrest and the deaths of several demonstrators (Birdsall and Nellis, 2002). On the other hand, some raw studies show some positive links between privatization and welfare. In particular, McKenzie and Mookherjee (2003) argue that the sale of state-owned enterprises brought positive welfare effects and that the poorest segments of the population appear to be relatively better off. In Argentina, they cite falling electricity prices that improved the welfare of all income deciles. They also report welfare gains in Bolivia from increased electricity access for all but the top income deciles. In Nicaragua, the authors argue that the value of gaining access to electricity was positive and of a larger magnitude for lower-income deciles who had relatively less access before privatization. Also, Galiani et al. (2005) designed tests that mapped water delivery to infant mortality in order to address concerns about quality after privatization. They show that Argentinean child mortality fell by 5 to 7 percent more in areas that privatized water services than in those that did not. The effect was larger in the poorest municipalities that privatized, where child mortality fell 24 percent. Privatization translated into 375 child deaths prevented per year. McKenzie and Mookherjee's study (2003) is, however, at most suggestive, as it has been soundly criticized for the weakness of the data used, identification problems, and analytical leaps and extrapolations (Saavedra, 2003).

This chapter studies the link between a privatized utility and household income in a developing country by taking advantage of a quasi-natural experiment that occurred in Peru as a result of the privatization of the state-owned telecommunications enterprise in the early 1990s. In particular, the privatization contract called for the privatized firm to install public telephones in 1,526 small rural towns distributed along the national territory in a random fashion. Using a household survey designed and performed by the authors on a representative sample of towns that received treatment until 2001, it was possible to study household income and welfare implications using both conventional regression analysis and matching methods. The basic premise is that telecommunication services reduce the gap in access to both formal and informal information. This reduction of informational gaps reduces the ability of the better-informed to extract rents from the less-informed, and thus helps enhance resource allocation and improve income and welfare among those living in more disadvantaged areas, in particular rural ones.[2] In fact, the existing empirical literature on the impact of rural telecommunications on income is scarce and far more suggestive than formal. Bayes (2001) argues that a village pay phone program in Bangladesh may be an example of how pragmatic policies can turn telephones into production goods. Services originating from telephones in villages may be more likely to deliver more benefits to the poor than to the non-poor. Matambalya and Wolf (2001) analyze the effect of information technologies on the performance of small and medium-sized enterprises and suggest that there may be no effect. Saunders, Wardford and Wellenius (1994) try to analyze the savings associated with the use of telecommunications services instead of alternative means of communication. Finally, the International Telecommunications Union (1998) reports a series of anecdotal material on the benefits of rural telecommunications.

Rural Telecommunications, Information, and Economic Outcomes

Even though telecommunications infrastructure has long been recognized as a key ingredient in promoting economic growth (Röller and Waverman, 2001), it has not been a central investment issue in many developing countries. Low-income countries account for only 6 percent of the world's telephone mainlines, which equals about 28 lines per 1,000 inhabitants. In contrast, high-income countries account for 52 percent of the world's telephone mainlines, or 585 lines per 1,000 inhabitants. Furthermore, a

---

[2] Developing countries tend to be more rural than is typically believed. For instance, when using a multidimensional definition of what rural is, it was found that 42 percent of individuals in Latin America live in such areas (De Ferranti et al., 2005).

comparison with rural areas makes the difference even more dramatic: in developing countries the rural mainline density is lower than one per 1,000 inhabitants (International Telecommunications Union, 2003). It has been claimed that under government control, the potential gains associated with access to telecommunications services are ignored, underestimated or simply unknown.[3] This lack of knowledge may be explained by several factors. First, telecommunications services are often considered a consumer good for the wealthy. Second, network externalities associated with telecommunications infrastructure are typically ignored. Third, while the empirical research has focused on the benefits of roads, transport, electricity and irrigation, little attention has been paid to the role of telecommunications. Fourth, the benefits of telecommunications infrastructure are often held to be positive axiomatically and thus, little is known about the size and distribution of those benefits, particularly in rural areas.

Rural telecommunications services constitute a crucial part of rural infrastructure since they provide the means for transferring information in a context where alternative means of obtaining information are less accessible. Advocates of this kind of infrastructure investment point out that the development of such infrastructure reduces information gaps, decreases the distance between economic agents and therefore reduces transaction costs. As a consequence, it enhances efficiency of resource allocation (Leff, 1984; Tschang et al., 2002; Andrew and Petkov, 2003). Information is a key component in enabling economic agents to make optimal decisions. It has been widely accepted, however, that most of the economic decisions are made under conditions of imperfect information; thus, decision makers may reduce their uncertainty through acquisition of additional information (Stigler, 1961; Stiglitz, 1985 and 2002). The ability to access and process information is recognized as a significant determinant of economic performance. In particular, productivity reflects not only how efficiently inputs are transformed, but also how well information is applied to resource allocation decisions (Allen, 1990; Babcock,

---

[3] There is convincing empirical evidence that, under government control, funds are not allocated on the basis of economic criteria (López-de-Silanes, 1997).

1990; Hubbard, 2003).[4] Information may be obtained from either formal or informal sources. Where formal or official information is limited or nonexistent, informal channels such as family and friends constitute an extremely important pathway of communication. Furthermore, recent research analyzes the role of social networks as a manner of obtaining information about job opportunities and explores its implications for the dynamics of employment (Durlauf, 2002; Calvó-Armengol and Jackson, 2004). In particular, the informal channel seems to be a non-negligible pathway to consider since people employ family networks to facilitate current and future transaction and credit flows. For instance, when asked to name their primary sources of information on a number of tasks related to cultivation and new technologies, the majority of farmers say they get information from family members (Godtland et al., 2004).

Data

As a result of the privatization of the state-owned telecommunications enterprise in 1994, the Peruvian government and Telefónica de España, the buying firm, agreed on a six-year investment schedule whereby the privatized firm was required to install and operate public pay telephones in 1,526 towns out of a list of about 40,000 rural municipalities. The basic characteristics of the eligible towns were that they did not have telecommunications services and were limited to a population of between 400 and 3,000 people. The towns were chosen randomly throughout the national territory.[5]

---

[4] There is a related body of literature on information diffusion and technology adoption in rural areas. Kebede et al. (1990) find that the likelihood of technology adoption increases with the level of education and access to information. Huffman and Mercier (1991) find that exposure to off-farm work increases the odds of adopting new technologies. Feather and Amacher (1994) find that a lack of information may be a reason the adoption of new practices has not occurred. Isham (2002) finds a positive link between adoption of farm technologies and the cumulative proportion of adopters, the presence of tribally based social affiliations, and the distance to local markets.

[5] Since monitoring was relatively lax, the authors examined whether the privatized company may have used particular criteria to choose the towns, such as the average income of the town, the density of the population, or potential linkages to larger areas, that may result

As part of the research for this chapter, a household survey was designed and implemented in 2002 on a representative sample of towns in rural areas in Southern Peru, a region that is characterized by extremely high levels of poverty.[6] In fact, this particular geographic area was chosen specifically because it is considered among the poorest in the country. The sample includes 1,000 rural households engaged in farm and non-farm activities, distributed proportionally between towns without any telecommunications service and towns with public telephones installed and operated by the privatized company. Ten households were randomly selected from each of 100 towns originally sampled. Thus, half of the towns had at least one public telephone installed by the privatized company in the most accessible part of the town, such as the municipal authority building or the main store in town. The other half, in which the lack of public telephone service was due primarily to a supply constraint instead of a demand constraint, was used as a control group.[7] The survey procedure followed a two-stage random sampling procedure and focused on the main demographic and housing characteristics of the household, as well as employment, farming activities, income, expenditures, availability of infrastructure, information and communication technologies, among other characteristics. Table 6.1 summarizes the characteristics of the variables used.

Table 6.2 shows that among surveyed households, more than 76 percent of the heads of household use the public telephone installed and

---

in sample selection bias. They did not find such evidence, since the distributions of the corresponding subsamples are not statistically different. There are two possible reasons for this. One, the rural pay phone investment requirement was a minuscule part of the total investment requirement of the privatized firm. Second, this type of investment may have been used by the company as a tool to increase goodwill and credibility in the face of necessary price increases.

[6] According to official figures, the poverty rates in 2001 for the four departments included in the survey were 75 percent (Cuzco), 44 percent (Arequipa), 78 percent (Puno), and 79 percent (Apurímac). Overall, the sampled area contained about 41 percent of all rural public telephones installed by Telefónica after privatization and comprises 25 percent of Peru's total rural population.

[7] People who live in towns without access to a public telephone travel many kilometers to reach a town that has the service. Other means of telecommunications, such as cellular telephones, are nonexistent.

Table 6.1 Summary Statistics

Variable	Obs	Mean	Std. Dev.	Min	Max
Total per-capita income (log)	1000	6.868	1.031	3.811	11.306
Total per-capita non-farm income (log)	761	6.496	1.352	3.062	10.300
Total per-capita farm income (log)	733	5.979	1.071	2.156	11.306
Family size	1000	4.895	2.026	1.000	14.000
Age of household head	991	44.259	14.052	20.000	90.000
Gender of household head (male)	1000	0.890	0.313	0.000	1.000
Years of education (average household)	1000	5.742	3.103	0.000	16.000
Squared years of education (average household)	1000	42.596	43.931	0.000	256.000
Spanish mother tongue (head)	1000	0.284	0.451	0.000	1.000
Work as dependent (head)	1000	0.494	0.500	0.000	1.000
Belongs to religious organization	991	0.269	0.444	0.000	1.000
Access to electricity	999	0.697	0.460	0.000	1.000
Time to the closest important town	1000	2.020	4.318	0.000	78.000
Town with public telephone	1000	0.500	0.500	0.000	1.000
Altitude (/1000 meters)	1000	3.125	0.775	0.163	3.978
Time to the closest public telephone	1000	0.872	1.176	0.000	5.000
Use of telephone services (head)	1000	0.766	0.424	0.000	1.000
Purpose of calls: Business	1000	0.101	0.301	0.000	1.000
Rural telephone expenditures (S/.)	1000	8.377	14.893	0.000	100.000
Apurimac	1000	0.281	0.450	0.000	1.000
Arequipa	1000	0.199	0.399	0.000	1.000
Cusco	1000	0.400	0.490	0.000	1.000
Puno	1000	0.120	0.325	0.000	1.000

operated by the privatized company. The usage is positively correlated with income. While around 65 percent of the bottom income group uses the town's public telephone, 88 percent of the top income group uses it. In terms of the expenditure on public telephone services, Table 6.2 shows that it varies from US$0.60 for the bottom income quartile to US$6 for the top income quartile. These expenditures represent 1.7 percent of the total household's income for the bottom quartile and 1.3 percent for the top quartile. In addition, the average number of telephone calls per month varies from 0.5 calls to 6.9 calls depending on income group. Again, households with higher incomes make more telephone calls. Furthermore, there appears to be a supply effect since the availability of a public telephone at the town level appears to have some impact on telephone usage. Among surveyed households, those

	Table 6.2	Use of Public Telephone, Travel Time and Direct and Indirect Expenditures

	HH's Income[1]	Use of Public Telephone[2]	Avg. Travel Time[3]	Average Call[4]	Direct Monthly Exp. on Phone[5]
Income Group					
I: Bottom 25%	35	65%	80	0.5	0.6
II	74	70%	64	0.9	1.0
III	147	84%	39	2.1	2.1
IV: Top 25%	463	88%	27	6.9	6.2
Type of Village					
With Phone	232	83%	7	3.0	2.8
Without Phone	127	71%	99	2.1	2.2
Total	180	77%	53	2.6	2.5

Source: Primary survey

Notes: All income figures are in dollars. The exchange rate employed is US$1 = 3.38S/ (World Bank, 2001).

[1] Refers to average monthly income of the household including both farm and non-farm income in U.S. dollars.
[2] Refers to the head of the household.
[3] Refers to average walking travel time to reach the nearest publicly accessible telephone, in minutes.
[4] Refers to average number of calls per month.
[5] Includes rates only.

from towns with an installed telephone have a higher usage rate than those households from towns without a telephone. This, perhaps, is a consequence of higher transaction costs, since travel time to reach the public phone is dramatically higher. Table 6.3 shows the one-way travel time to the nearest public telephone as a determinant of the rate of usage of telecommunications services. As expected, the longer the travel time to the nearest public telephone, the lower the usage. This implies that the higher the non-tariff cost to the nearest telephone, the lower the usage.

Regarding the main purpose of telephone calls, the survey reveals that households use telephones for both economic and social purposes. The most important reason for the use of a public telephone is to contact relatives (78 percent). The second most important reason is to do business (11 percent). Finally, the third most important reason is for emergencies (10 percent). It is important to recall that the percentage of business calls may be underestimated because it is common to observe that small rural farmers employ family networks to facilitate current

| Table 6.3 | One-Way Travel Time to the Nearest Public Telephone and Usage Rate | |

One Way Travel Time	Percentage of Sample in Category	Use of Public Telephone[1]
Within the village[2]	22	89%
Within 30 minutes distance	44	78%
Within one hour distance	8	72%
More than one hour distance	27	65%
Total	100	77%

[1] Refers to the head of the household.
[2] Refers to zero or negligible distance.

and future transactions and credit flows. This is particularly true in the Andean area (Cotlear, 1989; Mayer, 2002). Thus, it is not surprising that many telephone calls reported in the survey as being made "to contact relatives" are actually calls that have a business-related component (Godtland et al., 2004). On the other hand, there are a significant number of incoming calls whose main purpose is unknown. About 65 percent of the total traffic of rural public telephones is explained by incoming calls (OSIPTEL, 1999).

Regression Analysis

In the context of the discussion above, the aim of this chapter is to test whether rural telecommunications services improve the income of the household by helping reduce the gap in access to both formal and informal in-formation. To evaluate this, a simple empirical reduced form was estimated, and measures of total household income, farm, and non-farm per-capita income were linked with several telecommunications characteristics such as availability of telephones, distance to the nearest telephone, frequency of use and motive for using. When variables that may have potential reverse causality problems are included, an instrumental variables approach is also applied along with standard ordinary least squares methods. In particular, two variables that may be problematic in this respect are access to telephone, and telephone expenditures. Not only does more access to telephones help increase household income when, as argued above, the telephone is used as a business and information tool, but also higher income may allow more access to telephone use. Similarly, it is unclear whether telephone expenditures are conducive to higher household income or whether higher income leads to more telephone expenditures. The instruments employed are: (i) whether Spanish is the mother tongue in the household; (ii) whether the household belongs to a religious organization and (iii) whether the head of the household works as an employee in the formal sector.[8] Finally, all the regressions include fixed effects, which are applied at the departmental level.[9] The following specification is estimated:

$$y = \alpha + \beta \mathbf{H} + \lambda T + \varepsilon \tag{1}$$

where y is a measure of per-capita income; α is a constant; $\mathbf{H}$ is a vector of household characteristics; T is a series of variables associated with access to or use of a public rural telephone installed and operated by the privatized firm, β and λ their respective coefficients' vectors, and ε is an error term. In order to better understand the channels by which access to public telephone may impact households, this analysis uses three measures of per-capita income as dependent variables. The first measure consists of the total annual household per-capita income, regardless of source. The second consists of only farm per-capita income. The third consists of only non-farm per-capita income. The logic behind analyzing farm and non-farm income separately is consistent with recent research on the economics of rural households. Non-farm income serves as a consumption smoothing mechanism that helps counterbalance the cyclical nature of farm income. In fact, in the sample, the income share of non-farm activities is a high 53 percent, which is consistent with other studies on rural income (Escobal, 2002).[10] As such, it is also far more dependent

---

[8] Given the fact that these instruments may not be ideal, a complementary approach is also used by applying matching methods below.

[9] Non-fixed effects regressions were also tested and the findings are very similar.

[10] Per-capita expenditures were used instead of income measures. While very similar findings were obtained when compared with the total household income measure, expenditures do not allow for the separation of farm and non-farm activities, which as shown above, is particularly relevant in research related to rural areas. Empirical findings using expenditures are available upon request.

on outside linkages with adjacent regions—such as neighboring towns or nearby urban areas—if any. In a context where towns had lacked access to any telecommunications service prior to the installation of public telephones by Telefónica, inhabitants of poor rural villages will benefit the most from mechanisms that help them improve communication with other towns and villages; prior to the installation of those telephones, the only link to the outside world involved reaching the nearest village, typically on foot. As a result, chances are that demand possibilities will greatly increase.[11]

Figure 6.1 shows the kernel densities functions of these three income variables. As described above, these measures are regressed against a set of variables that have been classified in two groups. The first group contains a standard set of family characteristics: average years of schooling for household members, family size measured as the number of members in the household, age of household head, gender of household head (male), and walking time to the nearest town of similar size. The second group includes five variables related to access and usage of rural public telephones. The first is a dummy variable that captures the availability of rural telecommunications services. This dummy variable equals one when a town has a public rural telephone installed and operated by the privatized company. The second variable measures the distance, in hours, to the closest telephone service. The third is a dummy variable that equals one if the head of household uses the telephone service installed and operated by the privatized firm. The fourth variable is a dummy variable that equals one when the head of household reports business-related use of the telephone service. Finally, the fifth variable measures the intensity of use of telephone services as measured by the expenditure on telephone services.

Table 6.4 presents a basic set of estimates using the logarithm of the total per-capita income as the dependent variable. As expected, a positive and statistically significant link was found between total per-capita income and average years of schooling. A negative but sta-

---

[11] A similar example is provided in the case of rural roads (Jacoby, 2000; Escobal, 2002).

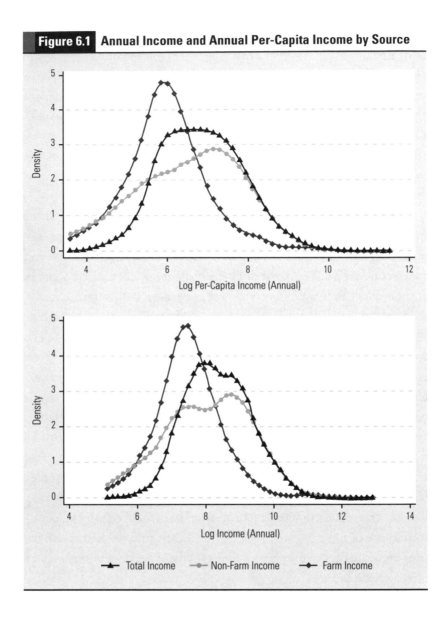

Figure 6.1 Annual Income and Annual Per-Capita Income by Source

tistically insignificant link was also found between the squared average years of education and income per capita. Family size yields a negative and statistically significant relationship. Similarly, age has a positive and significant link to household income. On the other hand, gender yields no significant link with total per-capita income. More important, with

Table 6.4 Access to Public Telephone in Rural Towns and Household Income

	OLS	OLS	OLS	IV	OLS	OLS	IV
	(1)	(2)	(3a)	(3b)	(4)	(5a)	(5b)
Family size	-0.1412***	-0.1443***	-0.1456***	-0.1486***	-0.1465***	-0.1594***	-0.1656***
	(0.012)	(0.012)	(0.012)	(0.013)	(0.012)	(0.011)	(0.014)
Age of household head	0.0044***	0.0051***	0.0056***	0.0064***	0.0056***	0.0040**	0.0034*
	(0.002)	(0.002)	(0.002)	(0.002)	(0.002)	(0.002)	(0.002)
Gender of household head	0.0744	0.0738	0.0268	-0.0154	0.0214	0.0305	0.0127
	(0.080)	(0.079)	(0.081)	(0.081)	(0.080)	(0.075)	(0.075)
Years of education (average household)	0.2001***	0.1886***	0.2088***	0.1866***	0.2104***	0.1509***	0.1238***
	(0.025)	(0.026)	(0.026)	(0.029)	(0.026)	(0.024)	(0.038)
Years of education sqrd. (average household)	-0.0020	-0.0015	-0.0021	-0.0013	-0.0023	-0.0009	-0.0002
	(0.002)	(0.002)	(0.002)	(0.002)	(0.002)	(0.002)	(0.002)
Time to the closest important town	-0.0137	-0.0079	-0.0108	-0.0123	-0.0097	-0.0123	-0.0133
	(0.009)	(0.010)	(0.009)	(0.009)	(0.009)	(0.008)	(0.008)
Town with public telephone after privatization	0.3079***						
	(0.049)						
Time to the closest public telephone		-0.1531***					
		(0.023)					
Use of telephone services (head)			0.1606***	0.4988**			
			(0.057)	(0.177)			
Purpose of calls: business					0.3642***		
					(0.081)		
Log. rural telephone expenditures						0.2425***	0.3406***
						(0.020)	(0.103)
Constant	5.9501***	6.3272***	5.9257***	5.8060***	5.9673***	6.1407***	6.2185***
	(0.170)	(0.175)	(0.177)	(0.185)	(0.174)	(0.159)	(0.171)
Obs	991	991	990	986	991	991	986
R^2	0.491	0.4974	0.475	0.459	0.481	0.543	0.531

All regressions include fixed effects. The dependent variable is log per-capita income (farm and non-farm). Standard errors are given in parentheses. The use of telephone services and rural telephone expenditures was instrumented. The instruments are whether Spanish is the mother tongue, whether the household belongs to a religious organization, and whether the head of household works as dependent. *** Significant at 1 percent; ** significant at 5 percent; * significant at 10 percent.

respect to the variables of interest, all the relevant variables were found to have the expected sign and are statistically significant at conventional levels.[12] In particular, the availability of a rural public telephone installed by the privatized firm in the town or village is associated with 30 percent higher per-capita income, since the sign of the coefficient is positive and statistically significant at 1 percent. This is shown in Column 1. Furthermore, a negative and statistically significant link was found between walking time to the nearest telephone service and total household per-capita income, as shown in Column 2 of the same table. The farther the telephone service is, the less likely it will be used, and as a consequence, the less the informational advantages of the service will be obtained. Additionally, the use of a rural public telephone service is associated with 16 percent higher per-capita income, or about 49 percent when correcting for endogeneity, as shown in Column 3a and Column 3b. Along the same lines, households that self-report business-related use of telephone services are associated with 36 percent higher total per-capita income. This is also shown in Table 6.4.[13] Finally, a positive and statistically significant link was found between expenditures on telephone service and total per-capita income. The findings suggest that an additional 10 percent of expenditure on public telephone usage is associated with a 2.4 percent increase in total per-capita income, or 3.3 percent when correcting for endogeneity. This is shown in Column 5a and Column 5b in the same table. In short, the measures associated with access and use of public telephones installed and operated by the privatized firm point towards the idea that such basic services have been conducive to increased household income in rural areas.

Table 6.5 repeats the above exercise but focuses on per-capita non-farm income as the dependent variable. Again, all the variables of interest were found to yield the expected signs and are statistically

---

[12] Since these variables are highly correlated, they are included separately as regressors. A principal components approach was also applied, which yields similar statistically significant results. These findings are available upon request.

[13] This estimate should be considered as a lower bound since many family calls include business components, as explained above.

Table 6.5 Access to Public Telephone in Rural Towns and Non–Farm Income

	OLS	OLS	OLS	IV	OLS	OLS	IV
	(1)	(2)	(3a)	(3b)	(4)	(5a)	(5b)
Family size	-0.1700***	-0.1722***	-0.1758***	-0.1831***	-0.1785***	-0.1952***	-0.2173***
	(0.021)	(0.021)	(0.021)	(0.023)	(0.021)	(0.020)	(0.024)
Age of household head	0.0021	0.0030	0.0036	0.0072**	0.0032	0.0018	0.0005
	(0.003)	(0.003)	(0.003)	(0.003)	(0.003)	(0.003)	(0.003)
Gender of household head	0.0504	0.0435	-0.0048	-0.1148	-0.0117	0.0020	-0.0410
	(0.124)	(0.124)	(0.125)	(0.130)	(0.123)	(0.122)	(0.127)
Years of education (average household)	0.2512***	0.2492***	0.2581***	0.2139***	0.2542***	0.2035***	0.1406**
	(0.041)	(0.042)	(0.041)	(0.046)	(0.042)	(0.039)	(0.057)
Years of education sqrd. (average household)	-0.0036	-0.0035	-0.0038	-0.0026	-0.0037	-0.0028	-0.0017
	(0.003)	(0.003)	(0.003)	(0.003)	(0.003)	(0.003)	(0.003)
Time to the closest important town	-0.0233	-0.0155	-0.0198	-0.0250	-0.0183	-0.0196	-0.0209
	(0.023)	(0.025)	(0.024)	(0.023)	(0.023)	(0.024)	(0.024)
Town with public telephone after privatization	0.3238***						
	(0.081)						
Time to the closest public telephone		-0.1241***					
		(0.040)					
Use of telephone services (head)			0.2235**	1.1990***			
			(0.095)	(0.308)			
Purpose of calls: business					0.4956***		
					(0.116)		
Log. rural telephone expenditures						0.2581***	0.5274***
						(0.0w29)	(0.152)
Constant	5.9328***	6.2311***	5.9441***	5.4502***	6.0509***	6.1979***	6.3591***
	(0.289)	(0.287)	(0.294)	(0.329)	(0.288)	(0.269)	(0.277)
Obs	754	754	754	752	754	753	752
R^2	0.424	0.4198	0.415	0.343	0.424	0.463	0.410

All regressions include fixed effects. The dependent variable is log per-capita non-farm income. Standard errors are given in parentheses. Use of telephone services and rural telephone expenditures were instrumented. The instruments are whether Spanish is the mother tongue, whether the household belongs to a religious organization, and whether the head of household works as dependent. *** Significant at 1 percent, ** significant at 5 percent, * significant at 10 percent.

significant at conventional levels. For instance, the availability of a rural public telephone is associated with 32 percent higher per-capita non-farm income, since the sign of the coefficient is positive and statistically significant at 1 percent. This is shown in Column 1. As before, a negative but statistically insignificant link was also found between walking time to the nearest public telephone and per-capita non-farm income, as seen in Column 2 of the same table. Regarding the measure of use, it was found that the use of a rural telecommunications service is associated with 22 percent higher per-capita non-farm income, 119 percent when correcting for endogeneity. These figures are shown in Column 3a and Column 3b, respectively. Again, households that self-report business-related use of a telephone service are associated with 25 percent higher per-capita non-farm income. Finally, a positive and significant link between expenditure on telephone service and per-capita non-farm income was found. The results suggest that an additional 10 percent of expenditure on public telephone use is associated with a 2.58 percent increase in per-capita non-farm income, or 5.27 when correcting for endogeneity. These findings are presented in Column 5a and Column 5b. In summary, the findings in this table show that access to telephone service is associated with an increase in non-farm income in poor rural towns. Additionally, notice that the non-farm income regressions yield a positive and statistically significant link with respect to average years of schooling, a negative but statistically insignificant link with years of education, a negative and statistically significant link with family size, a positive and statistically insignificant link with respect to age, and no significant link in the case of gender.

Table 6.6 repeats the same exercise as above but focuses on farm income. The availability of a rural public telephone installed by the privatized firm is associated with 13 percent higher per-capita farm income. This is shown in Column 1. While, as expected, a negative and statistically significant link was found between distance and per-capita farm income, a significant link in the case of telephone use was not obtained, as shown in Column 3a and Column 3b. Furthermore, the same occurs with households that self-report business-related use of telephone service. Finally, a positive and statistically significant link was observed between expenditure on

Table 6.6 Access to Public Telephone in Rural Towns and Farm Income

	OLS	OLS	OLS	IV	OLS	OLS	IV
	(1)	(2)	(3a)	(3b)	(4)	(5a)	(5b)
Family size	-0.1544***	-0.1562***	-0.1564***	-0.1592***	-0.1560***	-0.1644***	-0.1771***
	(0.016)	(0.016)	(0.016)	(0.020)	(0.016)	(0.016)	(0.019)
Age of household head	0.0070***	0.0072***	0.0077***	0.0087**	0.0076***	0.0068***	0.0059***
	(0.002)	(0.002)	(0.002)	(0.004)	(0.002)	(0.002)	(0.003)
Gender of household head	0.0522	0.0548	0.0352	0.0155	0.0377	0.0606	0.0873
	(0.113)	(0.113)	(0.114)	(0.128)	(0.114)	(0.107)	(0.109)
Years of education (average household)	0.1423***	0.1334***	0.1457***	0.1259*	0.1507***	0.0987***	0.0250
	(0.032)	(0.032)	(0.033)	(0.070)	(0.033)	(0.032)	(0.062)
Years of education sqrd. (average household)	-0.0053**	-0.0050**	-0.0053**	-0.0045	-0.0057**	-0.0039*	-0.0015
	(0.002)	(0.002)	(0.002)	(0.003)	(0.002)	(0.002)	(0.003)
Time to the closest important town	-0.0033	0.0000	-0.0020	-0.0030	-0.0012	-0.0038	-0.0068
	(0.006)	(0.006)	(0.006)	(0.007)	(0.006)	(0.006)	(0.006)
Town with public telephone after privatization	0.1336**						
	(0.061)						
Time to the closest public telephone		-0.0894***					
		(0.027)					
Use of telephone services (head)			0.0910	0.3641			
			(0.073)	(0.890)			
Purpose of calls: business					0.1101		
					(0.121)		
Log. rural telephone expenditures						0.1860***	0.4433**
						(0.191)	(0.019)
Constant	5.6661***	5.8974***	5.6039***	5.4747***	5.6242***	5.6915***	5.7537***
	(0.214)	(0.227)	(0.217)	(0.463)	(0.215)	(0.208)	(0.233)
Obs	729	729	729	729	729	729	729
R^2	0.328	0.333	0.326	0.293	0.325	0.3606	0.291

All regressions include fixed effects. The dependent variable is log per-capita farm income. Standard errors are given in parentheses. Use of telephone services and rural telephone expenditures were instrumented. The instruments are whether Spanish is the mother tongue, whether the household belongs to a religious organization, and whether the head of household works as dependent.
*** Significant at 1 percent, ** significant at 5 percent, * significant at 10 percent.

telephone service and per-capita farm income. An additional 10 percent of expenditure in public telecommunications is associated with a 1.86 percent increase in total per-capita farm income, or 4.43 when correcting for endogeneity. This is shown in Columns 5a and 5b of the same table.

This section has provided evidence that access to telephone services in poor, rural towns in Peru has helped increase per-capita household income. While such increases have appeared in both farm and non-farm channels, it is remarkable that the economic impact on non-farm channels is substantially larger than that on purely farm channels.[14] This finding is, in a way, not surprising when one bears in mind the theoretical literature on the role of telecommunications as a provider of information, particularly in small and isolated rural towns such as those considered in this study.

Propensity Scores Matching Methods

The soundness of the regression method approach is based on two assumptions. First, it hinges on the conjecture that the correct functional form of the outcome has been selected. Second, it implicitly assumes that it is possible to adequately control for potential differences between users and nonusers of rural public telephone services that may arise from the voluntary nature of participation. Unless these potential differences are properly accounted for, a comparison of outcomes would potentially yield biased estimates of the impact of rural public telephone access. As explained above, the use of an instrumental variables approach may not fully solve the potential endogeneity problem since, although the instruments considered are the best that could be found, they may not suffice to eliminate endogeneity. In fact, one could always argue that the instruments are not clearly correlated with use and are clearly uncorrelated with outcome (Jalan and Ravallion, 2003). In order to both minimize the potential bias in the impact estimate due to selection of observables, and employ a useful alternative method that avoids the need for using instrumental variables, matching methods were used to construct a statistical comparison group.

---

[14] Statistical tests on the difference of coefficients of farm and non-farm for any given regression always yield statistically significant results at 5 percent or higher.

Table 6.7 presents the estimates from the logit regression for both placement ("town with telephone") and usage ("head of household uses the telephone service"). The regressors comprise a wide range of household characteristics, dwelling characteristics, and geographical variables. In the case of variables placement (towns with telephone) prior to matching, the average estimated propensity score for treated and nontreated units were 0.6178 with a standard error of 0.1958, and 0.389 with a corresponding standard error of 0.2159. After matching, those numbers became 0.6140 with a standard error of 0.1930, and 0.6019 with a standard error of

Table 6.7 Logit Estimations for Placement and Usage

	Placement	Use
Family size	−0.0408	0.0298
	(0.038)	(0.025)
Age of household head	0.0109**	−0.0098***
	(0.005)	(0.003)
Gender of household head	−0.5852**	0.2516*
	(0.247)	(0.147)
Years of education (average household)	0.1074	0.1203**
	(0.088)	(0.058)
Years of education sqrd. (average household)	0.0014	−0.0010
	(0.006)	(0.005)
Time to the closest important town	0.1751**	0.0360
	(0.053)	(0.028)
Town with public telephone after privatization	0.1365	
	(0.104)	
Access to electricity	1.9457	0.3491***
	(0.183)	(0.109)
Spanish mother language (head)	0.4837**	0.5370***
	(0.239)	(0.181)
Altitude /1000	0.1881	
	(0.121)	
Constant	−2.9824	−0.7492
	(0.743)	(0.587)
Obs	990	990
LR chi^2	249.010	135.390
Prob > chi^2	0.000	0.000
Pseudo R^2	0.1814	0.125

The first dependent variable equals one if the town has a public telephone installed by the privatized firm, but is zero otherwise. The second dependent variable equals one if the head of the household uses the rural telephone service, but is zero otherwise. Standard errors are given in parentheses. All regressions include fixed effects.
*** Significant at 1 percent, ** significant at 5 percent, * significant at 10 percent.

0.1871, in the region of common support. Similarly, in the case of use, prior to matching the average estimated propensity scores for treated and nontreated units were 0.793 with a standard error of 0.1333, and 0.671 with a standard error of 0.1438. After matching, those numbers became 0.787 with a standard error of 0.1307, and 0.793 with a standard error of

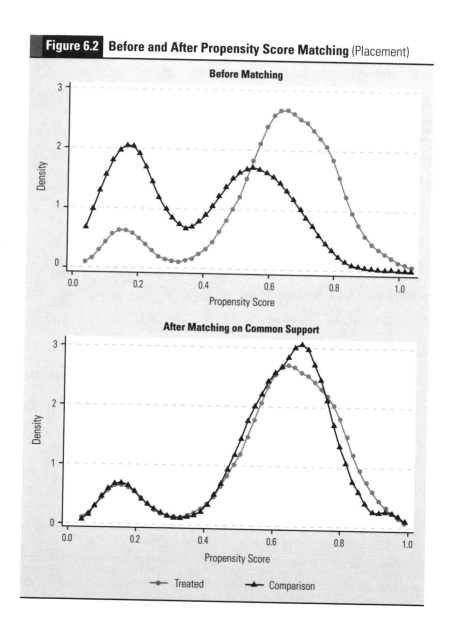

Figure 6.2 **Before and After Propensity Score Matching** (Placement)

0.1320, in the region of common support. Figures 6.2 and 6.3 show the kernel density of the estimated propensity scores for the two groups.

Table 6.8 reports the average treatment of the treated according to the kernel matching estimator. The results confirm a positive and significant link between access to telephone services and the measures

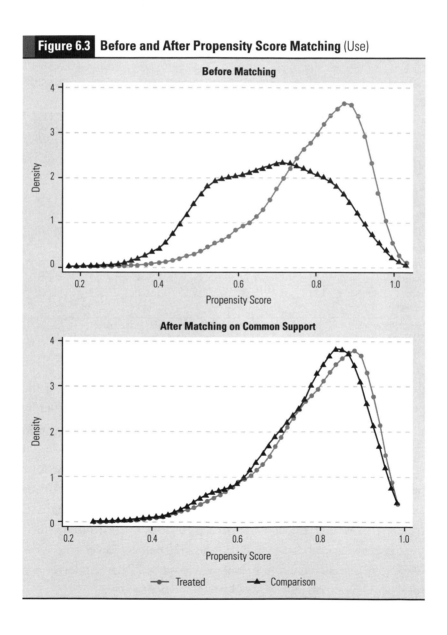

Figure 6.3 Before and After Propensity Score Matching (Use)

| Table 6.8 | Rural Public Telephone Usage: Average Treatment on the Treated |

	Outcomes					
	Non-Farm Per-Capita Income		Farm Per-Capita Income		Total Per-Capita Income	
	Mean	Std. Err.	Mean	Std. Err.	Mean	Std. Err.
Placement						
A. Unmatched						
Treated	1548	(110.6)	790	(185.5)	2338	(209.9)
Controls	640	(52.2)	509	(42.2)	1148	(64.5)
Treated-Controls	908	(122.3)***	281	(190.2)	1189	(219.6)***
B. Matched[1]						
Treated	1527	(108.6)	802	(188.8)	2329	(211.8)
Controls	1097	(19.5)	480	(4.6)	1581	(18.0)
Treated-Controls	429	(110.3)***	322	(188.9)*	747	(212.6)***
Use						
A. Unmatched						
Treated	1257	(79.0)	719	(123.6)	1976	(142.8)
Controls	561	(60.0)	421	(39.0)	982	(68.2)
Treated-Controls	696	(146.7)***	298	(224.7)	994	(261.2)***
B. Matched[1]						
Treated	1064	(75.3)	737	(134.1)	1794	(151.9)
Controls	833	(13.1)	433	(1.3)	1257	(13.4)
Treated-Controls (matched[1])	231	(76.4)**	304	(134.1)**	537	(152.5)***

[1] within the region of common support. Kernel Metric = P. Score; Kernel type: Epanechnikov (h=0.1).
*** Significant at 1 percent, ** significant at 5 percent, * significant at 10 percent.

of household income. Total per-capita household income among the population that is treated ("existence of public telephone installed by the privatized firm") would appear to be 32 percent lower without it. Furthermore, total per-capita non-farm income among the population that is treated would be around 28 percent lower otherwise. Similarly, total per-capita farm income among the population that is treated would be about 41 percent lower otherwise. Likewise, among the population that is treated ("uses rural telephone services installed by the privatized firm"), the total per-capita income would be about 30 percent lower otherwise, while total per-capita non-farm income would be 22 percent lower and total per-capita farm income would be 40 percent lower.

Conclusions

No doubt, privatization is under attack. Public opinion has turned against privatization and a large political backlash has developed, infused by accusations of corruption, abuse of market power, and neglect of the poor (Chong and López-de-Silanes, 2005). In a context where provision of basic services remains among the most pressing issues in developing countries, no companies have been more buffeted than those offering public utilities: water, electric and telephone services (Forero, 2005). This is particularly true in poor and rural areas where, according to the conventional wisdom, the welfare gains of privatization are rather questionable.

This chapter takes advantage of a quasi-natural experiment in Peru in which the privatized telecommunications company, Telefónica del Perú, was required by the government to randomly install and operate public pay phones in small and isolated rural towns along the national territory following privatization in 1994. Specially designed household survey data for a representative sample of rural towns allowed for the establishment of a link between access to telephone services and household income. Regardless of the income measurement, most characteristics related with access to public telephones installed and operated by the privatized firm were found to be positively linked with household income. Remarkably, such benefits occurred at both farm and non-farm income levels, with the latter being particularly crucial in rural areas because non-farm income primarily serves as an income-smoothing mechanism on which households tend to rely more and more (Chong, Hentschel and Saavedra, 2004). Not only do the findings hold when using instrumental variables, but they are also further confirmed when using propensity scores matching methods.

Critics of privatization may point to the fact that while the evidence presented in this chapter makes a positive case that increased access to telecommunications services in rural areas is conducive to higher household income, it does not provide convincing evidence on the benefits of privatization per se, since the rural investment studied in this paper was required by the government as part of the original privatization contract; as a consequence, it is not part of an investment strategy devised by the privatized company had pure laissez-faire been allowed. A more pragmatic

view of the findings is presented here. On the one hand, many governments include investment requirements in their privatization contracts
(Chong and López-de-Silanes, 2005). On the other hand, investment
requirements allow governments to leverage and direct resources from
privatized firms to regions where the private sector would not normally
get involved and where the government, because of lack of resources,
could not intervene either. Increased government tax collection may be
an issue, too. While the relevant question, which is beyond the scope of
this chapter, is whether investment requirements divert firm resources
from more productive uses elsewhere, it is also true that privatized firms
may have an incentive in participating in such investment deals because
they help promote goodwill on the part of the public and government at
a relatively low cost.[15]

---

[15] Recent private-public rural investment schemes appear to be good examples of the
private sector's interest in participating, at least in part, for signalling reasons (Wellenius,
Foster and Malmberg-Calvo, 2004).

CHAPTER SEVEN

Learning from an Incomplete Electricity Privatization Process in Rural Peru

Lorena Alcázar, Eduardo Nakasone and Máximo Torero

S ince 1990, Peru has embarked upon a drastic stabilization and structural reform process comprising a vast program of privatizing state-owned enterprises, including the main electricity and telephone utilities. In the electricity sector, the government approved the Law of Electric Concessions (DL 25844) in 1992, which separated power generation from transmission and electricity distribution, vertically unbundling the sector. Prior to the reforms, there were 12 state-owned distribution companies that were responsible for providing electricity service in Peru. In addition to vertical unbundling, the reforms included the reformulation of tariffs based on marginal costs, the introduction of a scheme of regulated and nonregulated markets, and the privatization of some key electricity assets above. The government also created the Supervisory Agency for Private Investment in Energy (OSINERG) to regulate tariffs and to control the quality and quantity of combustibles and service provision.[1]

Between 1994 and 1997, the government privatized 10 state-owned enterprises—five in electricity distribution and five in electricity generation—for a total of US$1.43 billion. As a result of this privatization process, 66 percent of installed generation capacity (MW) was administered by the private sector by 2005, and 60 percent of production (GWh) and 61 percent of billing came from private companies. In terms of transmission, 100 percent

---

[1] "What is OSINERG: Institutional Information." http://www.osinerg.gob.pe/osinerg/informa/qosinerg.jsp).

Table 7.1	Peruvian Electricity Sector Main Indicators	
General Indicators	**1993**	**2005**
Power capacity (MW)	4,282	6,201
Production (GWh)	14,678	25,510
Energy provided (GWh)	8,311	20,701
Number of clients	2,104,868	3,977,100
National electrification coefficient (%)	56.80%	78.10%
Distribution loss (%)	21.80%	8.41%
Investment (US$ millions)	174	394

Source: Ministry of Energy and Mines (MEM).

of transmission was managed by private companies by the same year, with US$122 million in billing. In sum, the private sector served 46 percent of the total number of clients in Peru by 2005 and distributed 71 percent of the energy, which accounted for 67 percent of distribution billing.[2]

Privatization in the distribution segment led to investments of US$838.9 million between 1994 and 2004, representing 61.27 percent of total investments reported in the same period.[3] As shown in Table 7.1, by the end of 2005 Peru had achieved a national electrification coefficient of 78.1 percent, up from the 56.8 percent reported in 1993, with generation capacity increasing by 67 percent during the same period. Electrification rates, however, still differ significantly among regions, in particular between the urban and rural sectors.

The incomplete privatization process has led to the existence of selected private provision areas while the rest of the country remains served by state-owned companies, but the National Rural Electrification Plan has promoted a broad range of initiatives since 1993 that target the poorest areas.

Unlike other experiences, the incomplete privatization of the electric sector in Peru provides a unique scenario by which to evaluate the impact of public versus private provision of public services within the same country.

---

[2] "Anuario Estadístico 2005." OSINERG—Gerencia Adjunta de Regulación Tarifaría División de Distribución Eléctrica, 2006.

[3] Appendix 1, Table A.1 of Alcázar, Nakasone and Torero (2007), the working paper on which this chapter is based, details the investments that took place by company.

This scenario makes it possible to compare differences in access, service quality, and other outcomes of the provision of electricity for the rural poor. In particular, the use of data collected through a specialized electricity and energy household rural survey allows a comparison of differences in welfare between people with private provision of electricity and people in regions where electric companies were not privatized.

Moreover, besides analyzing the impact on direct consumer welfare, this chapter also examines other indirect impacts such as the type of energy sources used by consumers in privatized versus nonprivatized areas, the effects of better quality of electric service on time devoted to nonagricultural activities, and the effects on time devoted to economic and noneconomic activities of rural households.

Overview of the Privatization of the Electric Sector in Peru and the Dynamics of Public and Private Operators
The Privatization Process

The privatization of the electric sector in Peru took place within a broader privatization process that started in 1991 with the Law for the Promotion of Private Investment in State Enterprises (DL 674). An ad hoc commission, COPRI, was created to conduct the whole process, and three special committees (CEPRIs) were formed to conduct specific processes within the electric sector (Electrolima, Electroperu, and Regional Electric Companies).

In addition, the electric privatization process was conceived as part of a broader reform of the electricity sector, introduced with the Law for Electric Concessions (1992). As a result, state-owned companies were spun off in that same year into several smaller companies, in the distribution, transmission, and generation segments, to fulfill the vertical unbundling clauses of the law and to provide some basis for yardstick competition. Four companies were created to serve the distribution market in Lima, and other regional companies were designed to cover broader geographical areas, as shown in Table 7.2.

As detailed in Figure 7.1 and Table 7.3, the privatization of the electric sector started in 1994 with the distribution companies EDELNOR and Edelsur (later Luz del Sur), both with concession areas in Metropolitan Lima, and then continued with Ede Chancay and Ede Cañete in 1995 and 1996, respectively, which served the provinces of Lima and which were acquired by the same economic groups controlling EDELNOR (Ede Chancay) and

Table 7.2 Public Company Spin-Offs			
Public Company Spin-offs	**Segment**	**Concession Area (only for distribution)**	**CEPRI**
		Electrolima	
EDELNOR	Distribution	Metropolitan north of Lima, Callao, Huaura, Barranca, Huaral and Oyón	CEPRI Electrolima
EDELSUR	Distribution	Metropolitan south of Lima	
Ede Chancay	Distribution	Chancay (Huacho, Huaral and Super)	
Ede Cañete	Distribution	Cañete	
Edegel	Generation		
		Electroperu	
Egenor	Generation		CEPRI
Cahua	Generation		Electroperu
Etevensa	Generation		
EEPSA	Generation		
		Regional companies	
Electro Sur Medio	Distribution	Ica, and part of Huancavelica and Ayacucho	CEPRI Regional
Electro Norte Medio	Distribution	La Libertad, Ancash and part of Cajamarca	Companies
Electro Centro	Distribution	Huánuco, Pasco, Junín, and part of Huancavelica and Ayacucho	
Electro Norte	Distribution	Lambayeque, Cajamarca and Amazonas	
Electro Noroeste	Distribution	Tumbes and Piura	
Electro Sur	Distribution	Tacna and Moquegua	
Electro Sur Oeste	Distribution	Arequipa	
Electro Sur Este	Distribution	Puno, Cuzco, Apurimac and Madre de Dios	
Egemsa	Generation		
Egasa	Generation		
Egesur	Generation		

Source: COPRI.

Figure 7.1 Privatization Timeline for Distribution Electric Companies

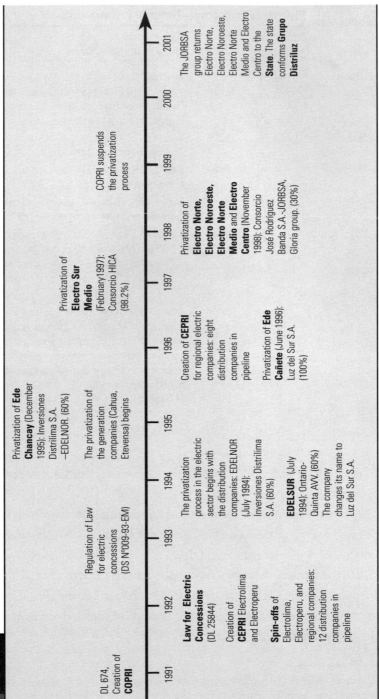

Source: COPRI.

Table 7.3	Privatized Companies, Terms of Privatization		
Company	Date	Price (US$m)	Terms (controlling stake)
EDELNOR	Jul-94	209.3	Sale of 60% stake to Inversiones Distrilima (controlled by Endesa (Spain), Chilectra, and Enersis (Chile) and Cosapi (Peru)). Cash. No investment commitments.
Luz del Sur	Jul-94	406.9	Sale of 60% to the Ontario Quinta consortium. Cash. No investment commitments.
Ede Chancay	Dec-95	10.5	Sale of 60% to EDELNOR. Cash. No investment commitments.
Ede Cañete	Jun-96	8.6	Sale of 100% to Luz del Sur. Cash. No investment commitments.
Electro Sur Medio	Feb-97	51.28	Sale of 98.2% to HICA Consortium. 40% cash (20% up front and difference in eight years); 50% through investment commitments in rural electrification; 10% to be shared with workers.
Electro Norte Medio		67.88	
Electro Centro	Nov-98	32.69	Sale of 30% to JORBSA (10% up front and difference in 12 years). An option to acquire an additional 30% of the company was included. No investment commitments but with the requirement of operating rural electrification projects handled by the DEP under their area of influence.
Electro Noroeste		22.88	
Electro Norte		22.12	

Source: COPRI.

Luz del Sur (Ede Cañete). As a result, the Lima area is basically 100 percent covered by privatized companies.

The privatization of the electricity distribution companies in Lima consisted mainly of the sale of a 60 percent stake to a private strategic operator, while 10 percent of capital was offered to the companies' workers, and the remaining 30 percent was sold on the stock market (with the exception of Ede Cañete, which was sold entirely to Luz del Sur). Clearly, the main objective of these processes was to maximize the proceeds of the sale for the Peruvian government.[4]

---

[4] In all cases, the selection process consisted of two main stages. In the first stage, bidders had to prequalify based on financial indicators and credentials; sometimes no previous ex-

The CEPRI for the regional electricity companies was created in 1996 initially to manage the privatization of the eight state-owned regional distribution companies: Electro Sur Medio, Electro Norte Medio, Electro Centro, Electro Norte, Electro Noroeste, Electro Sur, Electro Sur Oeste, and Electro Sur Este.

The privatization of Electro Sur Medio took place in 1997, followed in 1998 by the joint privatization of Electro Norte, Electro Norte Medio, Electro Noroeste, and Electro Centro. Electro Sur Medio was awarded to the HICA consortium of the Argentine IATA and the Peruvian C. Tizón, Amauta Industrial, S&Z Consultores Asociados, and Constructores Vásquez Espinoza S.A., while Electro Norte, Electro Norte Medio, Electro Noroeste, and Electro Centro were awarded to the Peruvian firm J. Rodríguez Banda S.A.-JORBSA (Gloria group).

The privatization scheme designed for the regional electricity distribution companies differed from the way the process was handled in Lima. With the objective of promoting electrification in unserved areas, the model included investment commitments of US$25.64 million to expand the electrification frontier in the case of Electro Sur Medio, representing 50 percent of the total payment, and no investment commitments but an obligation to serve any potential demand under a delimited area of influence in the case of the group of regional companies in the north and center.

In 1999, during the process of privatizing Egasa, the state-owned company controlling the southern region of Arequipa's main hydroelectric plant, protests interrupted the privatization process. Electro Sur, Electro Suroeste, and Electro Sureste therefore remained in public hands. The privatization process suffered an additional setback when Electro Norte, Electro Noroeste, Electro Norte Medio, and Electro Centro were returned to the state by the JORBSA group in December 2001. The government then began to design a new privatization process for the same companies in 2002, which is why the companies have maintained their private

---

perience was required. Then, competition in the second stage was based on the economic offers of qualified bidders. The competitive factor used was the largest payment over a price preestablished by COPRI (between US$8.2 and US$129 million, depending on the size and importance of the company).

structure, operating together under the name "Distriluz," outside of the common legal framework for public companies. For instance, their investments do not enter into the national public investment system, nor are they subject to the government procurement laws governing the other public companies; Distriluz is discussed further below.

Provision of Electricity in Rural Peru

The state embarks upon rural electrification projects in order to provide energy to people who are not served by the existing electrical system because of the distance and lack of accessibility of their dwellings. Most of the time, it is not profitable for the existing companies to supply energy to these areas because of the huge investment needed, so public intervention is essential.

To address the problem of rural electrification, a special division within the Ministry of Energy—the Executive Directorate of Projects (DEP)—was created in 1993 to handle energy projects and extend the electrical frontier. The DEP received the mandate to create, fund, and implement the National Rural Electrification Plan (and to update it annually during a 10-year timeframe).

In 2002, a specific law promoting rural electrification was passed (Law 27744) creating a fund for rural electrification under the administration of the DEP and defining criteria for the National Rural Electrification Plan. However, the law was inapplicable given its reported contradictions with the Decentralization Law (Law 27783) and the Organic Law of Regional Governments (Law 27867), and because it was too general in various aspects. Recently, a new general law of rural electrification (Law 28749-2006) was enacted to replace the previous one. This new law is more extensive, more clearly defines the Rural Electrification System, and extends the sources of financing (for example, from 2 to 4 percent of the profits of the electrical companies, and consumer contributions of 2/1,000 of 1 UIT[5] per MWh consumed). Also, the law involves a

---

[5] The Unidad Administrativa Tributaria (UIT) is a tax-related reference unit. As of September 2006, the value of one UIT is around US$1,045.

systematic effort to organize rural electrification works through the National Rural Electrification Plan, considers the role of ADINELSA, and includes an entire section on promotion of private investment in rural electrification.

The electrification rate for Peru's rural sector was estimated at 32 percent in 2002, up from a reported 5 percent in 1992. Rural areas adjacent to urban centers have benefited from the expansion of the electrification frontier by distribution concessionaires following the reforms introduced in the sector since 1992, but some of the poorest and most remote areas in rural Peru have also found access to the service as a result of a broader range of initiatives fueled by other sources, mainly the National Rural Electrification Plan. The expansion of the electrification frontier by concessionaires and by all these alternatives, however, has not been part of an organized and systematic plan, and information on the importance of the various modalities of service provision and their results is therefore limited.

From its creation in 1993 to 2004, the DEP invested US$552 million in rural electrification, providing 4.9 million inhabitants with access to electricity. The works are financed with its own resources (including income from privatizations) and with external contributions. The DEP engages directly in the construction of infrastructure and then transfers the operation and maintenance assets to several different actors, including distribution companies (when the project is under their concession area) or to ADINELSA, a public holding created to find an operator for built infrastructure (when the project is outside of concession areas). Since it became operational in 1998, ADINELSA has mainly transferred networks and systems to the distribution companies that currently operate under Distriluz (Electro Noroeste, Electro Norte, Electro Norte Medio, and Electro Centro) because of commitments in the privatization contract and has also promoted the creation of municipal or community-based units to operate other infrastructure. These operations are usually not profitable, so ADINELSA provides the necessary subsidies to the operator (from a fund based on contributions from concessionaries).

Public efforts to promote rural electrification have not only consisted of investing in rural electrification projects, but also of establishing a new

| Table 7.4 | FOSE Subsidy Classifications | | |

Users	Sector	Tariff reduction for consumers =< 30 kWh/month	Tariff reduction for consumers > 30 kWh/month up to 100 kWh/month
Interconnected Systems	Urban	25% energy charge	7.5 kWh/month
	Urban–rural & rural	50% energy charge	15 kWh/month
Isolated Systems	Urban	50% energy charge	15 kWh/month
	Urban–rural & rural	62.5% energy charge	18.75 kWh/month

Source: Law No. 28307 28.07.04.

tariff structure to make electricity affordable for the poor. In August 2001, the Peruvian Congress passed legislation to establish a "social tariff" for electricity consumption. The enacted law created the Social Compensation Fund (FOSE), a cross-subsidy that benefits final users who consume less than 100 kWh per month by providing them with a discount that varies depending on predetermined ranges of consumption (see Table 7.4 for details on the subsidy classifications). This fund is financed by a 3 percent tax on final users who consume more than 100 kWh per month. The FOSE differentiates tariffs according to both the quantity of kilowatt hours consumed and the area of consumption, providing more benefits to rural consumers. Although the FOSE was originally envisioned to be applied for only three years, Congress indefinitely extended the application of the subsidy in July 2004.

Public versus Private Provision of Electricity

Of the 22 distribution companies currently regulated by OSINERG, nine were privatized (of which five remain private and four were returned to the state after three years of private management); seven have always been public; and six are new private companies created after the reform of the sector. Map 7.1 and Table 7.5 present the concession areas, showing a scenario of multiple operating models in the provision of electricity in Peru, which forms the basis of this evaluation of the impact of privatization.

Map 7.1 | Map of Main Distribution Companies' Concessions

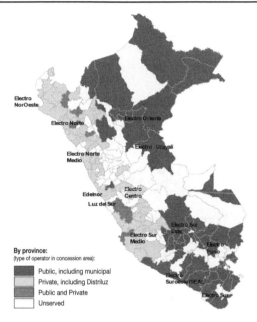

By province:
(type of operator in concession area):

- Public, including municipal
- Private, including Distriluz
- Public and Private
- Unserved

Source: MEM.

Currently, electricity in Peru is provided through two main actors: (i) the traditional spun-off state-owned distribution companies and (ii) privatized distribution companies. In addition, several new private distribution companies have been established under operating models which include schemes of subsidized electricity infrastructure investment, defined and carried out through governmental and not-for-profit institutions, which are later transferred for their operation to existing distribution companies or newly created units at local governments. Municipal or community-based operating units have also appeared as the result of the decentralization laws.

The penetration of both state-owned and privatized distribution companies has been shaped significantly by the structure of the electricity sector as conceived by the Law for Electric Concessions, which grants zonal concessions to the main distribution companies. In essence, there is one operator per concession area to achieve economies of scale, and a

Table 7.5 Distribution Companies

Company	Property	Number of Customers	Concession Areas
Consorcio eléctrico de Villacuri S.A. (Coelvisa)	Private	644	Lima, Ica and Huánuco
Electro Paramonga	Private	5,687	Paramonga
Electro Utcubamba (Emseusa)	Private	4,902	Utcubamba
Electro Pangoa	Private	1,033	Pangoa
Electro Rioja	Private	4,087	San Martin
Electro Tocache	Private	7,114	Tocache
Ede Cañete	Privatized	25,978	Cañete
EDELNOR	Privatized	912,186	Metropolitan north of Lima, Callao and the provinces: Huaura, Barranca, Huaral and Oyón
Edechancay	Privatized		Chancay (Huacho, Huaral and Supe)
Electro Sur Medio	Privatized	123,311	Ica, and parts of Huancavelica and Ayacucho
Luz del Sur S.A.	Privatized	719,651	Metropolitan south of Lima
Chavimochic	Public	3,316	La Libertad
Electro Oriente	Public	126,581	Loreto, San Martin
Electro Puno	Public	115,656	Puno
Electrosur	Public	95,896	Tacna and Moquegua
Electro Sur Este	Public	228,696	Puno, Cuzco, Apurimac and Madre de Dios
Electro Ucayali	Public	41,811	Ucayali
Electro Centro	Mixed**	364,957	Huánuco, Pasco, Junín, and parts of Huancavelica and Ayacucho
Electro Norte Medio-Hidrandina	Mixed**	396,563	La Libertad, Ancash and part of Cajamarca
Electro Noroeste	Mixed**	228,753	Tumbes and Piura
Electro Norte	Mixed**	218,346	Lambayeque, Cajamarca and Amazonas
Sociedad Eléctrica del Sur Oeste	Mixed**	234,477	Arequipa
Total		3,859,645	

Source: OSINERG and MEM.
**Privatized and returned to the state.

requirement to provide service to whomever requires it within the concession area and to facilitate installations for other operators that require it (within a one-year period). Expansions in these areas have been achieved mainly through connecting new users to the national grid.

Other actors, however, such as newly created private companies and municipal entities, operate under different incentive schemes. Private companies have mainly been created to supply the electricity demand of selected agricultural and other business-related areas, while municipal and community-based initiatives have been developed as a result of national or municipal endeavors of various types—mainly to serve the most remote areas. Since the enactment of the 1992 law, only six new private companies have been created and are almost exclusively oriented to the business sector (not residential). In contrast, there is evidence of a significant number of small municipal and community-based initiatives. Most of these access gains have been achieved through the construction of small and isolated power systems.

Classifying State-Owned Distribution Companies and Privatized Companies

This chapter classifies as public distribution companies those that have always been in public hands. As shown in Map 7.1, they are the following: Electro Sur, Electro Sureste, Electro Puno, Electro Ucayali, and Electro Oriente. Through concessions, these companies cover more than 450,000 square kilometers in Peru's coastal, Andean, and jungle regions and, as of December 2004, served 843,000 clients (representing 22 percent of the total). These companies have expanded to rural areas within their concession areas. In addition, the DEP has transferred several projects to them within the framework of the National Rural Electrification Plan.

Private companies, on the other hand, need to be classified into two groups:

i) Electro Norte, Electro Norte Medio, Electro Noroeste, and Electro Centro (together Distriluz), which were privatized and remained in private hands for three years before being returned to the state in 2001. The Distriluz companies have concession areas in Peru's northern coast and the central Andean region, covering 180,200 km and serving 1.2 million clients by the end of 2004 (31 percent of total clients in the country).

ii) Companies that were privatized and are still private, that is, EDELNOR, Luz del Sur, Ede Cañete, and Electro Sur Medio. These companies cover 40,584 square kilometers in concession areas and serve 46 percent of total clients.

The "Distriluz" companies constitute a peculiar case that deserves further analysis to determine its classification as private or public. As noted above, these public companies were privatized in 1998. While these companies were in private hands—the Rodríguez Banda Group (JORSA–Gloria Group)[6]—important changes took place, resulting in a significant reduction of electricity losses in distribution,[7] for example, and a reversal of negative financial balances. A few years later in 2000, however, the Gloria Group argued that it could not afford to fulfill its investment commitments. After negotiations, an arrangement was made and the Gloria Group returned control of the four companies to the state. They were registered as the "Distriluz" group in 2001 (having a single directory and sharing policies as they did when in private hands). It is formally a mixed company, but private firms have only very minor participation.[8] Nonetheless, Distriluz is treated by the regulatory agency OSINERG and the tax agency as a private firm and is not required to comply with other public firms' regulations (Law 674), such as registration and approval of investments by the National Public Investment System (SNIP) or observance of the State Contracting and Procurement Law or the Public Indebtedness System. Furthermore, as Distriluz directors and personnel are hired through private contracts, its labor regime is based on Law 728, which regulates the activity of private firms, and the companies manage their own budgets.

This arrangement was possible because of a special exclusion norm (COPRI 363) that allowed the Distriluz group to maintain a private regime because it was supposed to return to private hands within a short

---

[6] And other minor private partners, most of whom are still shareholders.

[7] Distribution energy losses were 18.68 percent in 1998 and were reduced to 9.88 percent by the end of 2001.

[8] Electro Noroeste and Electro Centro are 100 percent public, while Electro Norte is 99.99 percent public; in the case of Hidrandina, 5.3 percent of total shares are in private hands.

Table 7.6	Number of Customers in Rural Areas (2005)		
Empresas	Interconnected Systems	Isolated Systems	Total
Edecañete	5,735	71	5,806
Electro Oriente	0	2,824	2,824
Electro Puno	4,589	1	4,590
Electro Sureste	128,668	0	128,668
Electro Sur Medio	9,610	1,152	10,762
Electro Ucayali	654	0	654
Electro Centro	170,118	5,642	175,760
Electro Noroeste	28,039	301	28,340
Electro Norte	12,779	25,653	38,432
Electro Sur	9,678	0	9,678
Emseusa		4006	4,006
Hidrandina	41,240	9,023	50,263
Seal	15,775	5,011	20,786
Total	468,185	79,100	547,286

Source: OSINERG.

period. However, the exclusion norm has been renewed every year since the return of Distriluz to public ownership. This situation is very peculiar, as it is the only case of a public company that functions under private market rules.

Under their privatization contract, the Distriluz companies (unlike other private distribution firms) were required to provide electricity service not only in their concession areas but also, if required, in a larger area of influence (mostly rural), which was defined in the contract. As a result, these companies operate in various rural areas (Table 7.6). In addition, and as shown in Table 7.7, the companies have also received several electricity distribution networks built up by the DEP under the auspices of the National Rural Electrification Plan, for operation and maintenance purposes.

In the Lima area, EDELNOR and Luz del Sur have reached almost 100 percent coverage of their concession areas, including some rural areas. In addition, both companies have expanded their influence to nearby rural areas in some provinces of Lima (Table 7.6). No transfers from DEP/ADI-NELSA have been reported.

Table 7.7	DEP and Adinelsa Rural Projects' Transfers to Distribution Companies	
	Number of rural projects transfered by	
Company	DEP	ADINELSA
Electro Sur Medio	16	n.a.
Coelvisa	n.a.	2
Electro Tocache	n.a.	2
Electro Norte Medio	39	n.a.
Electro Norte	32	38
Electro Centro	42	28
Electro Noroeste	24	8
Electro Sur	14	n.a.
Electro Suroeste	44	n.a.
Electro Sureste	74	n.a.
Electro Oriente	46	10
Electro Ucayali	16	n.a.
Electro Puno	20	n.a.
Municipalities	32	38

Source: DEP and ADINELSA.

In the immediate southern environs of Lima, Electro Sur Medio has contributed to the electrification of rural areas such as Huancavelica and Ica as part of its investment commitment (US$25.64 million). In addition, the DEP has transferred a few rural projects to the company for operation and management.

Finally, municipal and community-based electricity operations have been reported to exist in various areas across the country. Some of these operations have their origins in the National Rural Electrification Plan carried out by the DEP, while others originate from other initiatives such as the external cooperation mandates of international financial institutions. Given that in most of these cases the scope of the distribution service is under the 500 kW limit required to be a concession, it is difficult to identify them extensively. Furthermore, based on the survey under study,[9] their coverage is still limited (covering only about 3 percent of clients).

In summary, this scenario of multiple operating models providing electricity in rural Peru provides a unique opportunity to evaluate differ-

---

[9] National Survey of Rural Energy Demand.

ences in households' welfare resulting from private versus public provision of the service.

Data

One of the major characteristics of the privatization processes is a lack of detailed pre- and post-privatization information on the privatized firms. An advantage of the research undertaken for this chapter, however, is access to a representative survey of electricity and energy use in rural Peru. Specifically, the sample universe for the survey includes all communities in Peru with a thousand or fewer households, whether rural-dispersed, semirural or peri-urban. This survey, conducted during June and July of 2005 in the 24 departments of Peru, includes two components: population centers (communities) in each dominion of the study (446), and 6,690 households.

The survey is additionally representative of seven subregional divisions: Costa Norte (Northern Coast); Costa Centro (Central Coast); Costa Sur (Southern Coast); Sierra Norte (Northern Andean Region); Sierra Centro (Central Andean Region); Sierra Sur (Southern Andean Region) and Selva (Amazon region). The sample size is distributed according to Table 7.8.

In addition to including the traditional questions found on a Living Standards Measurement Study survey (which provide sufficient variables

Table 7.8	Distribution of the Sample Size		
Dominion	Sample Conglomerates	Sample Houses	Expected Standard Deviation (cv)
Costa Norte	64	960	0.032
Costa Centro	64	960	0.050
Costa Sur	48	720	0.038
Sierra Norte	66	990	0.021
Sierra Centro	68	1,020	0.029
Sierra Sur	68	1,020	0.024
Selva	68	1,020	0.022
Total	446	6,690	

to use in the matching procedure), this survey contains specific modules on electricity and energy consumption, which make it possible to answer the main questions of this study. Specifically, the survey includes information on households' sources of energy, and it collects detailed information on the provision of electricity for households connected to the grid. For these households, the survey collects information on the name of the distribution company (crucial in identifying areas where privatized companies are operating) and the year in which the household was connected to the grid, details on consumption and the last three electric bills, number of hours in a day households had access to service, and quality of service provision.

Based on the information obtained from the survey, and following the previous classification of private and public distribution companies, 24 percent of the households are served by private distribution companies; 15 percent are served by public companies; 3 percent are served by other models including regional, municipal, and communal operations; and the rest do not have access to electricity. It is also important to note that an important percentage of the households served by private distribution companies—16 percent—is served by the Distriluz group's regional distribution companies, which, as noted above, are no longer private but still retain some private elements.

Based on survey information, Map 7.2 shows the geographic location of the private and public operators and the number of observations. Private operators are located mainly in the coastal and central Andean regions.

Table 7.9 shows the means of the main variables included in the database. As can be seen, most households without access to electricity are worse off than households with access to electricity, either through private or public sources. Households without access to electricity are less educated, have more members, have less access to other infrastructures, work mainly in farm activities, and have significantly lower per-capita income and expenditure.

As seen in Table 7.10, a comparison of households with private access versus public access to electricity also shows significant differences, although in this case the differences are smaller and the direction is not

Map 7.2 **Map of Public and Private Operators by Location**

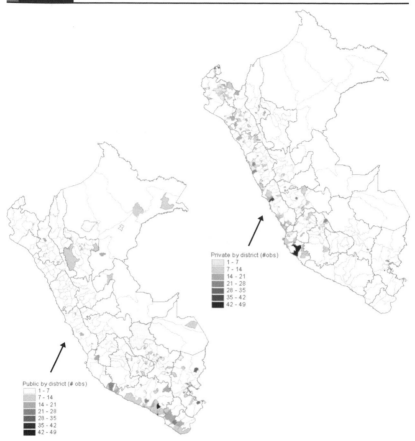

Source: National Survey of Rural Energy Demand.

clear (i.e., they are not consistently better off in one case or the other). For example, households with private access have a statistically significant higher number of members, the household head is older, there are more female household heads, they are less likely to be indigenous, they work fewer hours, and they concentrate more on non-farm activities. Therefore, it is necessary to make these two groups more comparable in order to analyze the potential impacts of private versus public provision of electricity. The following section details the methodology developed to make the groups comparable.

Table 7.9	Distribution of sample by electricity company	
Company	**OBS**	**Departments**
Private companies		
Electro Norte Medio (Hidrandina)	341	Ancash, Cajamarca, La Liberta
Electro Noroeste	266	Piura, Tumbes
Electro Sur Medio	264	Ayacucho, Huancavelica, Ica
Electro Centro	243	Pasco, Junin, Huanuco, Huancavelica, Ayacucho
Electro Norte	204	Lambayeque, Cajamarca, Amazonas
EDELNOR	115	Lima
Ede Cañete	96	Lima
Luz del Sur	65	Lima
Sociedad Minera Colquirrumi	15	Cajamarca
Cmte Electrificacion Chugur	5	Cajamarca
Total private[1]	1614	
Total private[2]	560	
Public companies		
Seal	400	Arequipa
Electro Sureste	249	Madre de Dios, Cusco, Apurimac
Electro Puno	171	Puno
Electro Sur	117	Moquegua, Tacna
Municipalities, local governments, and others	90	Various
Electro Oriente	86	Loreto
ADINELSA	58	Lima, Ica
Autodema	15	Arequipa
Electro San Gaban	15	Puno
Electro Shitariyacu	14	San Martin
Electro Ucayali	11	Ucayali
Total public[1]	1226	
Total public[2]	2280	

Source: national survey of rural energy demand.
Note: The first four companies listed make up the Distriluz group.
[1] Considering Distriluz as private.
[2] Considering Distriluz as public.

Statistical and Econometric Methodology

This section compares differences in the set of dependent variables between households in privatized areas and a comparison group consisting of a sample of similar households in communities (*centros poblados*) where

Table 7.10 Summary Statistics of Main Variables under Study at the Household Level

	Without electricity	With electricity		Diff is significant?		
	(1)	Public (2)	Private (3)	(1) & (2)	(1) & (3)	(2) & (3)
Household characteristics						
Age of household (HH) head	48.08	48.91	50.91	***		***
Years of education of HH head	5.00	7.07	6.97	***	***	***
Female head of household	0.13	0.14	0.18	***	***	***
Indigenous HH head	0.32	0.45	0.19	***	*	***
Household size (members)	4.49	4.19	4.36	***	***	**
Proportion 0–5 years old	0.12	0.09	0.08	***	***	
Proportion 6–13 years old	0.19	0.18	0.18	*	**	
Proportion 14–60 years old	0.54	0.59	0.58	***	***	
Proportion 60 years or older	0.15	0.14	0.16			*
Household with pipeline water	0.17	0.54	0.59	***	***	***
Other sources of water access	0.23	0.16	0.13	***	***	**
Water from river, lake, etc.	0.27	0.19	0.20	***	***	
Town characteristics						
Population (in thousands)	350.69	1228.41	1336.68	***	***	**
Public phone in town	0.09	0.54	0.46	***	***	***
Secondary school in town	0.13	0.56	0.52	***	***	**
Paved road in town	0.26	0.63	0.61	***	***	
Town in coast	0.35	0.39	0.59	**	***	***
Town in highlands	0.46	0.47	0.36		***	***
Town in jungle	0.18	0.14	0.05	***	***	***

(continued on next page)

Table 7.10 Summary Statistics of Main Variables under Study at the Household Level (contd.)

	Without electricity	With electricity		Diff is significant?		
	(1)	Public (2)	Private (3)	(1) & (2)	(1) & (3)	(2) & (3)
Time allocation and welfare						
Hours of work – HH	12.18	12.54	10.90	*	***	***
Hours of work – HH (incl chores)	19.17	19.16	17.27		***	***
Proportion of non-farm hours of work – HH	0.23	0.39	0.49	***	***	***
Proportion of non-farm hours of work – HH (incl chores)	0.14	0.25	0.29	***	***	***
Hours of leisure (TV and radio)	3.24	4.17	4.42	***	***	**
Hours of leisure (TV, radio, socializing, and others)	6.98	7.71	9.16	***	***	***
Proportion of non-farm income	0.31	0.33	0.40		***	***
Per-capita expenditure	136.47	198.24	217.25	***	***	***
Energy						
Number of sources of electricity	4.00	3.09	2.83	**	***	***
Expenditure on electricity (% of expenditure on energy)	N/A	0.47	0.48			
Expenditure on electricity (% of total expenditure)	N/A	0.05	0.05			
Number of sources of electricity	N/A	3.09	2.83			***
Price per kW	N/A	1.08	1.00			
Number of monthly failures (30+ min)	N/A	1.32	1.30			
Often dimming in electric service	N/A	0.13	0.11			**
Monthly hours of blackouts	N/A	5.62	4.59			**

* Significant at a 10% level of confidence; ** Significant at a 5% level of confidence; *** Significant at a 1% level of confidence.

privatization has not yet taken place, as shown in Map 7.2. To construct the required comparison group, a two-step matching process was followed and is detailed in Figure 7.2.

Figure 7.2 **Two-step Matching**

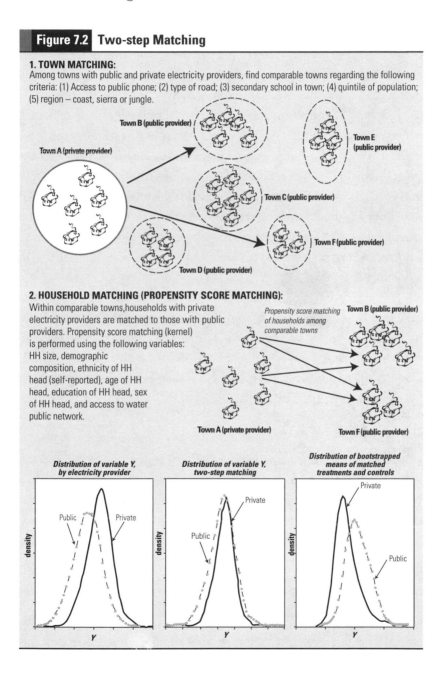

1. TOWN MATCHING:
Among towns with public and private electricity providers, find comparable towns regarding the following criteria: (1) Access to public phone; (2) type of road; (3) secondary school in town; (4) quintile of population; (5) region – coast, sierra or jungle.

Town B (public provider)

Town A (private provider)

Town E (public provider)

Town C (public provider)

Town F (public provider)

Town D (public provider)

2. HOUSEHOLD MATCHING (PROPENSITY SCORE MATCHING):
Within comparable towns, households with private electricity providers are matched to those with public providers. Propensity score matching (kernel) is performed using the following variables: HH size, demographic composition, ethnicity of HH head (self-reported), age of HH head, education of HH head, sex of HH head, and access to water public network.

Propensity score matching of households among comparable towns

Town B (public provider)

Town A (private provider)

Town F (public provider)

Distribution of variable Y, by electricity provider

density

Public

Private

Y

Distribution of variable Y, two-step matching

density

Public

Private

Y

Distribution of bootstrapped means of matched treatments and controls

density

Private

Public

Y

The process first consists of a nonparametric town matching and a household matching based on propensity scores with the distribution of bootstrapped means of matched treatments and controls. With respect to town matching, the control town and treatment town pairs are estimated nonparametrically using cell means based on access to public telephone, type of road, secondary school in town, quintiles of population, and region (coast, sierra, or jungle). The resulting matching yielded a distribution of treated and control towns that were approximately symmetrically distributed.

Second, household matching is based on propensity scores of households within each control and treatment town group. The framework serving as a guideline for the empirical analysis is the Roy-Rubin model (Roy, 1951; Rubin, 1974). Inference on the impact of a treatment (privatization in this case) on the outcome of an individual involves speculation about how this individual would have responded had he or she not received the treatment. Therefore, the objective will be to capture what happened to a household in a privatized area versus what happened to a household in a nonprivatized area.

A binary assignment indicator, D, is defined, indicating whether a household unit actually was affected by the program ($D = 1$, meaning the household is located in a community where electricity is provided by a private firm) or not ($D = 0$) (Hujer and Wellner, 2000; Lechner, 2000). The treatment effect of each household is then defined as the difference between its potential outcomes:

$$\Delta = Y^T - Y^C \tag{1}$$

where Y will be the income (welfare effect) of the household and the superscripts refer to the treatment group (T) and the control group (C) where the treatment is an area in which the electric company (distribution) was privatized.

The fundamental problem of evaluating this household treatment effect arises because the observed outcome for each household unit is given by:

$$Y = D \cdot Y^T + (1 - D) \cdot Y^C \tag{2}$$

Unfortunately, Y^T and Y^C for the same household unit can never be observed simultaneously. The unobservable component in Equation (2) is called the counterfactual outcome, so that for households that participated in the measure ($D = 1$), Y^C is the counterfactual outcome, and for those that did not, it is Y^T. Therefore, there will never be an opportunity to estimate households' gains with confidence. As a result, it is necessary to concentrate on the population average of gains from treatments, that is, the average treatment effect on the treated:

$$E[\Delta \setminus D = 1] = E(Y^T \setminus D = 1) - E(Y^C \setminus D = 1) \tag{3}$$

As Hujer and Wellner (2000) note, this parameter provides an answer to the following question: "What is the expected or mean outcome gain for individuals who received treatment as opposed to a hypothetical situation where they do not receive it?" This question focuses directly on actual participating units, so that it determines the realized gross gain from the program and can be compared with its costs. This will help decide whether the program is a success or not (Heckman, Ichimura and Todd, 1997, 1998; Heckman, LaLonde and Smith, 1999).

The second term on the right side of Equation (3) is unobservable since it describes the hypothetical outcome without treatment for those units that received treatment. If the condition:

$$E(Y^C \setminus D = 1) = E(Y^C \setminus D = 0) \tag{4}$$

holds, nonparticipants can be used as an adequate control group. This identifying assumption is likely to hold only in social experiments, where the key concept is randomized assignment of household units into treatment and control groups. In nonexperimental data, as in this study, Equation (4) will normally not hold. The use of nonparticipants as a control group will therefore lead to a selection bias. Heckman and Hotz (1989) point out that selection might occur on observables and unobservables.

An approach is here attempted that estimates the unobserved counterfactual term using the observed outcome information obtained from the nonparticipants and taking into account selection on observables as well as selection on unobservables (see Hujer and Caliendo, 2000, for a detailed description).

To develop the control group, the propensity score matching approach is used. The basic idea underlying the matching approach is to search within a large group of nonparticipants in the matched towns to find those individual units that remain similar to the participants in all relevant pretreatment characteristics. A variable summarizing all these relevant characteristics is estimated as the probability of receiving based on a probit model.[10] Households in the treatment group are matched to relevant controls as a function of differences in their propensity scores. From among the different approaches as to how to match households in both groups (i.e., one-to-one, radial, "k" nearest neighbors, local linear regression, etc.), kernel matching with a quadratic function (Epanechnikov) was chosen. Under the latter methodology, each household of the treatment group is matched to a weighted average of all available controls. For each treatment, a set of weights is estimated for every control as a function of the difference between its propensity score and that of the control. Once a suitable counterfactual for every treatment observation has been found, the differences in the outcomes between the well-selected and adequate control group and the participants can then be attributed to the program.

Finally, bootstrapping is used to recover the empirical distribution of the treatments and the controls. The significance level of the differences is then computed after 10,000 iterations following Davidson and MacKinnon (1999).

Empirical Results

Theoretically, propensity score matching should be able to balance characteristics not related to treatment, but it is prone to affecting the outcome variables between treatment and control groups. Thus, if all the relevant characteristics remain the same, with the only difference being in the treatment, it can be inferred that discrepancies in the outcome

---

[10] Several procedures for matching the propensity score can be used. A good review can be found in Heckman, Ichimura and Todd (1998).

variables between both groups can be attributed to the program under assessment.

The results suggest that the proposed two-step propensity score matching procedure performs better, finding more rigorous counterfactuals for households with private provision of electricity. This improvement was tested by analyzing how well different matching methods balance the differences in a set of observable characteristics between households with public and private provision of electricity. To simplify, one-to-one matching procedure of households in the treatment and control groups was performed with three alternative procedures: (i) traditional propensity score matching considering household characteristics, (ii) traditional propensity score matching considering household and town characteristics in a single probit equation, and (iii) the proposed two-step estimator. This made it possible to observe not only the differences in the outcome variables (as in the case of the kernel matching), but also the differences in other observable characteristics of each treatment and its selected control household.[11] Our method provides better balancing between both groups and, overall, achieves satisfactory comparability.

The transfer from the public to the private sector (Vickers and Yarrow, 1988) necessarily implies a change in the relationships between those responsible for the firm's decisions and the beneficiaries of the profit flows (the social view and the agency view). In theory, the transfer of property rights leads to a different structure of management incentives, causing changes in managerial behavior, company performance, and quality of service in terms of access and use. In this chapter we concentrate on changes in the quality of service and on their respective impacts on a household's welfare. Specifically, the main objective of this paper is to assess the two main effects of the provision of electricity through a private provider: first, the effects on electric service quality, and second, whether this improvement in quality results in an improvement in a household's welfare through its effects on either changes in total hours of work or on time allocation between farm and non-farm activities.

---

[11] The results are presented in Appendix C of Alcázar, Nakasone and Torero (2007).

Changes in the Quality of Electricity Service

It is expected that a change in private provision, with different management structure incentives, should significantly improve the quality of the service. Specifically, the traditional literature normally mentions that privatization will have a significant impact on investment, access, quality of service, and a realignment of tariffs. In the Peruvian case, as shown above, there was a significant improvement in investment as a result of the privatization. Similarly, the coefficient of electrification increased significantly during the past decade. For example, currently 100 percent of EDELNOR customers (83 percent of whom belong to the poorer segments of Lima's inhabitants) have electricity. EDELNOR's investments have added to the network 225,000 customers in approximately 500 communities. Similarly, a significant realignment of tariffs was expected, so that they would reflect real costs, resulting in losses in the consumer surplus of households. As mentioned above, however, in 2001 a "social tariff" for electricity consumption was established (FOSE—Fondo de Compensación Social Eléctrica).

Finally, with respect to the quality of service, it is expected that consumers will face fewer service failures (less dimming in electric services, fewer hours of blackouts, and a lower number of failures), have more hours of electricity, use fewer sources of energy, and increase their consumption of electricity.

Table 7.11 presents the results of the electricity performance indicators at the household level after the two-step matching procedure. As expected, households that use distribution from privatized providers have better quality of electricity provision. Specifically, households report less dimming, a smaller number of monthly failures, and lower monthly hours of blackouts. This implies that they have a better quality of provision and a subsequent reduction in costs. Similarly, households with private provision have significantly higher expenditures on electricity, both as share of their total expenditure and also as share of their expenditure on energy sources. This result, together with the fact that prices are lower, implies that there is a clear increase in the amount of electricity consumption by these households.

Table 7.11	Matching Results on Performance Indicators					
	Mean				[95% Conf	
	Treat	Control	ATT	S.E.	Interval]	
Expenditure on electricity (% of expenditure on energy)	0.48	0.46	0.02	0.02	−0.01	0.05
Expenditure on electricity (% of total expenditure)	0.05	0.05	0.00	0.00	−0.01	0.01
Number of sources of electricity	2.83	3.19	−0.36	0.07	−0.49	−0.22**
Price per kW	1.04	1.14	−0.10	0.07	−0.36	−0.05**
Number of monthly failures (30+ min)	1.41	1.53	−0.12	0.16	−0.38	0.25
Often dimming in electric service	0.12	0.16	−0.04	0.01	−0.10	−0.01**
Monthly hours of blackouts[1]	4.72	5.66	−0.94	0.64	−2.24	0.24

Note: Standard errors are based on 10,000 iterations of the two-step matching procedure. See Appendix B for a description of the variables.
[1] The two-stage PS matching difference was significant at 10 percent.
** Denotes significance at a 95% level of confidence.

Additionally, and as expected, there is a significant reduction in the number of sources of energy used by households linked to private providers. Households can access energy from 14 possible sources, but given improvements in the quality and hours of electricity provision, they reduce their overall sources of energy from 3.169 to 2.84.[12]

The next question to ask is whether these improvements in the provision of electric service have an impact on a household's welfare, or what may be called potential indirect impacts.

Impacts of Better Quality of Electric Service on Hours of Work and on Time Allocation for Farm and Non-Farm Activities

The aggregate-level links between poverty and rural infrastructure have been studied by several authors, but among the most important of these

---

[12] The options are (1) electricity, (2) kerosene, (3) candles, (4) dry cell batteries, (5) car batteries, (6) liquefied petroleum gas, (7) solar home system, (8) firewood, (9) animal dung, (10) crop residues, (11) electric generator set, (12) charcoal, (13) coal, and (14) others. Additional details are provided in Appendix B.

works are Lipton and Ravallion (1995), Jiménez (1995), and Van de Walle (1996), in addition to those cited above.[13]

In order to further analyze the effects of public infrastructure, and specifically the effects of the improvement of the quality of electricity due to the presence of a private distributor, it is necessary to distinguish between direct and indirect effects. The former occur when an increase in access to electricity is accompanied by an increase in production, shifting the production frontier and marginal cost curve, and also increasing the rate of return for private investment in rural activities. The latter take place when access to more or higher quality electricity permits a reduction in the transaction costs that small producers face when they integrate into the supply and factor markets. These lower transaction costs change the structure of relative prices significantly for the producer, stimulating changes as transitions occur in the allocation of the labor force between agricultural and nonagricultural uses. The latter effects are what this subsection attempts to measure.

There are three possible channels through which these indirect impacts on income may be affected by access to infrastructure and, in this specific case, to better quality of electricity provision. On the one hand, there is the impact of changes on the proportion of working hours allocated to different activities. Specifically, shifts in labor devoted to agricultural and nonagricultural activities are analyzed. The hypothesis is that access to better quality of electricity leads to greater opportunities for non-farm work activities. On the other hand, the second channel captures the effect of changes in the household's total working hours as a result of longer hours of access to electricity. Finally, there is scope for increases in rural households' market

[13] For a specific infrastructure impact case (like the role of rural roads, telephones, or access to electricity on poverty alleviation) the literature is very broad and includes works such as Howe and Richards (1984), Binswanger, Khandker, and Rosenzweig (1993), Jacoby (1998), and Lebo and Schelling (2001), among others. Recently, Renkow, Hallstrom, and Karanja (2003) estimate the fixed transaction costs (those not dependent on commercialized volume) that impede access to product markets by subsistence farmers in Kenya. These authors estimate that high transaction costs are equivalent to a value-added tax of approximately 15 percent, illustrating the opportunities to raise producer welfare with effective infrastructure investments. Smith et al. (2001) show that the rehabilitation of roads in Uganda increases labor opportunities in the service sector.

efficiency through increases in their purchasing power. Along these lines, the third channel captures changes based on returns to labor (that is, hourly wages) allocated to agricultural and nonagricultural activities. Specifically in the case of agricultural activities, this will be directly related to prices of their products. This study concentrates on the first two effects, which have a clearer relationship with better quality and hours of electricity.

Table 7.12 provides results for labor mobility and the allocation of labor between farm and non-farm activities. Better quality of electricity has a significant and positive effect on labor mobility to non-farm activities. Specifically, there is an increase in non-farm activities in the three indicators presented. Households with private provision of electricity allocate around 10 percent more of their working time to non-farm activities. Specifically, without including chores, they expend 50 percent of their time on average on non-farm activities, while households with public access expend an average of 40 percent of their time on non-farm activities. This proportion increases to 52 percent when the analysis is conducted at the individual level rather than the household level. The rural non-farm sector has developed as a major source of employment, and it seems that there

Table 7.12 Indirect Impacts of Better Access and Better Quality of Electricity

| | Mean | | | | [95% Confidence | |
	Treatment	Control	ATT	S.E.	Interval]	
Hours of work – HH	10.66	12.34	–1.68	0.35	–2.45	–1.09**
Hours of work – HH (incl chores)	17.00	19.18	–2.18	0.31	–2.94	–1.75**
Proportion of non-farm hours of work – HH	0.50	0.40	0.10	0.03	0.05	0.16**
Proportion of non-farm hours of work – HH (incl chores)	0.29	0.25	0.04	0.02	0.01	0.08**
Hours of work – individual (incl chores)	5.34	6.19	–0.86	0.18	–1.25	–0.53**
Hours of work – individual	8.50	9.68	–1.18	0.13	–1.55	–1.03**
Proportion of non-farm hours of work – individual	0.52	0.43	0.09	0.02	0.04	0.13**
Proportion of non-farm hours of work – individual (incl chores)	0.28	0.25	0.03	0.02	0.00	0.06

Note: Standard errors are based on 10,000 iterations of the three-step matching procedure. See Appendix B for a description of the variables.
** Denotes significance at a 95% level of confidence.

is a positive association between this kind of development and a better quality and duration of the provision of electricity. The non-farm activities that have developed so far, however, appear to be mainly in the tertiary sector, probably as a result of rural electrification. The secondary sector remains insignificant as a source of rural employment and income.

With respect to the total working hours, propensity score matching was used to formally assess the impact of private provision of electricity on total household hours worked per average day, as was the case with the share of farm and non-farm activities. The results show that households with access to electricity from private providers work an average of 1.7 hours less and per capita around 1.18 hours less. A possible explanation is that men and women allocate their time to a variety of economic and noneconomic activities with important implications for their income, health, leisure, and overall livelihood and well-being. As a result, better access to electricity and non-farm activities could allow them to increase their efficiency, reduce their total work burden, increase their leisure time, or earn more income using the same number of working hours.

When the time allocation of the household head and spouse between treatment and control households (see Table 7.13) is examined in detail, differences in time allocation are concentrated in three main activities. First, as mentioned above, there is a significant increase in time allocated to non-farm activities or more time allocated by the control on farm activities (4.34 percent); second, the treatment group spends more time on handcrafts and shop tending (1.62 percent) and other leisure activities (2.87 percent). This result is consistent with communities with more hours of electricity and with better quality of service, which allows them to assign part of their time to non-farm-related activities and to have better choices in leisure activities.

Finally, Table 7.14 presents some global measures of welfare. Consistent with the previous results, per-capita expenditure is not statistically different between both types of households. This could be because major income changes occur as a result of having or not having access, and in this analysis both groups of households already have access to electricity. Moreover, households with better provision of electricity work fewer total hours because they can compensate for foregone hours of farm labor with

| Table 7.13 | Differences in Time Allocation between Treatment and Control Households[1] (Head of household + spouse[2]) |

	Hours		Distribution		
	Control	Treatment	Control	Treatment	Difference
Sleeping	16.13	16.25	33.61%	33.85%	0.24%
Bathing/grooming	0.90	1.13***	1.87%	2.36%	0.49%
Cooking	2.96	3.11	6.16%	6.48%	0.32%
Farming, gardening, animal husbandry, fishing	7.12	5.04***	14.83%	10.50%	−4.34%
Income earning act, such as handicrafts, tending shop, etc.	4.43	5.21**	9.22%	10.85%	1.62%
Eating	2.84	2.95	5.91%	6.15%	0.24%
Processing food	0.27	0.22	0.57%	0.45%	−0.12%
Water fetching and fuel collection	0.56	0.52	1.17%	1.09%	−0.08%
Laundry, house cleaning	2.31	2.19	4.82%	4.57%	−0.25%
Repairing clothes, basket equipment, tools, etc.	0.65	0.45***	1.36%	0.94%	−0.42%
Religious practices (praying, reading Bible, etc.)	0.37	0.34	0.77%	0.71%	−0.06%
Reading, studying	0.43	0.53	0.90%	1.10%	0.20%
Watching TV, listening to radio, resting	4.74	4.58	9.88%	9.54%	−0.34%
Visiting neigbors, socializing	1.57	1.85*	3.28%	3.85%	0.57%
Other leisure activities	1.52	2.90***	3.16%	6.03%	2.87%
Shopping	1.19	0.70***	2.48%	1.46%	−1.02%
Other	0.00	0.04	0.00%	0.09%	0.09%
Total	48.00	48.00	100.00%	100.00%	0.00%

[1] Nearest neighbor (caliper=0.01) two-step propensity score matching.
[2] Sample has been limited to households with available information for both head of household and spouse.
* Significant at a 90% level of confidence; ** Significant at a 95% level of confidence; *** Significant at a 99% level of confidence.

| Table 7.14 | Indirect Impacts of Better Access and Better Quality of Electricity on Total Welfare |

	Mean				[95% Confidence	
	Treatment	Control	ATT	S.E.	Interval]	
Per-capita expenditure	219.25	228.10	−8.85	8.75	−31.11	3.70
Proportion of non-farm income	0.41	0.32	0.09	0.03	0.02	0.14**
Hours of leisure (TV and radio)	4.55	4.60	−0.05	0.17	−0.42	0.26
Hours of leisure (TV, radio, socializing, and others)	9.41	7.86	1.54	0.29	0.97	2.10**

Note: Standard errors are based on 10,000 iterations of the two–step matching procedure. See Appendix B for a description of the variables.
** Denotes significance at a 95% level of confidence.

activities in the non-farm sector, which pay higher salaries. Therefore, the results seem to indicate that the quality of electricity matters in increasing the availability of non-farm activities and therefore the distribution of working hours between farm, non-farm, and leisure activities. This result is clear when the proportion of non-farm income is analyzed. Non-farm income accounts for 41 percent of the total income of households with private provision of electricity and 32 percent of the income of households with public provision. In addition, households with private provision spent about 1.5 hours more on leisure activities. In summary, better quality of electricity implies a more efficient allocation of rural households' time, which allows them to have more time for recreation.

Conclusions

The passage of the Law of Electric Concessions (DL 25844) in November 1992 set the stage for a comprehensive reform of Peru's electric sector. Reinforcing the subsidiary role of the state, the new operating and institutional frameworks sought to maximize efficiency and enhance competition in all electricity activities. Those frameworks, while promoting the system's interconnectivity, provided for a vertical unbundling of the sector that segmented the power generation, transmission, and distribution activities, and defined free competition and regulated markets.

Of the three large segments—generation, transmission, and distribution—the generation segment operates under perfect competition and can be conducted by private and public enterprises. Within the transmission segment, the main network system is operated by a private operator under a 30-year concession scheme with the government. Other secondary transmission lines are also mostly private, but the government participates with some investments.

The distribution segment consists of a mixture of public and private providers and is mostly regulated, given its characteristics as a natural monopoly. Currently, based on the number of clients served, electricity distribution is 47 percent private. The presence of private and public providers offered a unique opportunity to evaluate the impact of public versus private provision within the same country.

The results can be summarized as follows: first, there is a significant improvement in the quality of the provision of electricity when distribution firms are managed by the private sector. This result is consistent with solid work that supports the proposition that privatization improves the operating and financial performance of firms (Galal et al., 1994; La Porta and López-de-Silanes, 1999; and the studies summarized in D'Souza and Megginson, 1999).

Second, improvements in the quality and supply of electricity provision yield some efficiency gains in terms of the time allocation of the working labor force that can be directly linked to the use of electricity. Rural households under private provision of electricity had more opportunities to work in non-farm activities, and as a result, the share of non-farm activities increased, indicating both a substitution effect and a potential price effect. The substitution effect implies a reduction of hours spent on farm activities in favor of non-farm activities and the price effect implies that households will receive higher salaries and therefore will need to work fewer hours in total. As a result, the increase in time spent on non-farm activities was accompanied by a reduction of hours spent on farm activities and an increase in hours spent on leisure.

Appendix A Private Investments (thousands of US$)

Private Companies	1994	1995	1996	1997	1998	1999	2000	2001	2002	2003	2004
Electrical partnership Villacuri S.A.C (Coelvisa)						281	856	36,591	491	180	568
EDELNOR	9,021	23,889	53,642	56,848	30,900	41,037	36,000	40,011	30,160	22,213	18,722
Electro Pangoa							24				
Electro Sur Medio	1,234*	5,788*	10,318*	3,330	4,169	2,487	3,356	1,847	935	586	236
Ede Cañete							222	239	509	458	1,588
Electro Paramonga							13	2			
Electro Tocache						17	160				
Electro Utcubamba	12	9	9	37	133				85		
Luz del Sur	19,854	34,628	44,516	39,342	41,203	28,805	26,900	29,419	32,841	24,802	30,507
Public Companies											
Electro Centro	7,337	11,849	17,145	6,522	1,285**	1,190**	20,337**	2,232**			
Electro Noroeste	10,700	10,705	12,108	19,868	11,467**	11,192**	4,478**	1,605**			
Electro Norte medio	4,819	16,692	4,796	1,239	1,196**	n.a.	28,761**	4,799**			
Electro Norte	8,134	2,493	10,362	4,692	4,179**	2,379**	1,878**	1,970**			
Electro Oriente	1,205	677	9,952	5,902	9,298	8,135	2,339	7,706	2,338	851	
Electro Puno							1,161	16	482	1,208	204
Electro Sureste	6,264	48,943	17,612	22,375	26,387	16,793	2,314	2,541	2,628	2,700	2,588
Electro Ucayali						1,122	1,363	559	337	93	
Electro Sur	1,211	998	3,856	2,857	1,957	1,093	1,293	999	1,537	1,279	1,150
Chavimochic							351				
Electrical services Rioja							5		1		28
SEAL	3,621	6,737	8,959	8,477	44,526	6,931	7,262	3,845	5,549	4,033	4,131
Total	73,400	163,399	193,278	171,461	176,576	121,499	139,206	134,381	77,893	57,552	60,573
Private investments	28,875	58,517	98,170	99,529	90,239	87,425	123,118	118,715	65,021	48,239	21,114
Percentage of private investments	39.3%	35.8%	50.8%	58.0%	51.1%	72.0%	88.4%	88.3%	83.5%	83.8%	34.9%

Source: MEM
*Property of the state; **Private property.

Appendix B Description of Variables

Variable	Description
Welfare	
Household per-capita expenditure	Includes expenditures on energy sources (electricity, kerosene, candles, dry cell batteries, car batteries, liquefied petroleum gas, solar home system, firewood, animal dung, crop residues, electric generator set, charcoal, coal and others); food; expenditures for water, telephone and transportation; home maintenance and repair; products of personal hygiene; recreation activities; health care, education, transport expenditures, clothing and shoes; and furniture and cooking utensils. Variable estimated in Metropolitan Lima soles, June 2005.
Total agricultural income	Includes income from self-employment in agriculture, livestock and fisheries; and salaried work. Salaried work was considered agricultural when the individual reported working in two-digit ISIC codes 1, 2 and 5. Around 15 percent of individuals declaring salaried income did not report the activity they were working in. To avoid misrepresentation due to missing data, households with any member declaring salaried income but lacking information on economic activity were not considered in the estimations.
Total nonagricultural income	Includes income from business and salaried work. Salaried work was considered agricultural when the individual reported working in two-digit ISIC codes 1, 2 and 5. Around 15 percent of individuals declaring salaried income did not report the activity they were working in. To avoid misrepresentation due to missing data, households with any member declaring salaried income but lacking information on economic activity were not considered in the estimations.
Proportion of non-agricultural income	Ratio of total nonagricultural income to nonagricultural income plus agricultural income.
Daily hours of work (1)	Sum of daily hours of work in agricultural and nonagricultural activities of head of household and spouse. Variable is only calculated for married households (i.e. with information for *both* head and spouse). Hours of agricultural work include farming, gardening, animal grazing, fishing, etc. Nonagricultural work includes employment in shop, production of handicrafts and others; processing food; repairing clothes, basket, machineries, equipment, etc.
Daily hours of work (2)	Sum of daily hours of work in household chores, agricultural activities and nonagricultural activities of head of household and spouse. Variable is only calculated for married households (i.e. with information for *both* head and spouse). Household chores include preparing meal, water fetching, washing clothes, house cleaning, and shopping. Agricultural work includes farming, gardening, animal grazing, fishing, etc. Nonagricultural work includes employment in shop, production of handicrafts and others; processing food; repairing clothes, basket, machineries, equipment, etc.

Appendix B	Description of Variables (continued)
Variable	**Description**
% of nonagricultural hours of work (1)	Daily hours of nonagricultural work as a proportion of daily hours of work (agricultural and nonagricultural hours of work). Variable is only calculated for married households (i.e. with information for both head and spouse).
% of nonagricultural hours of work (2)	Daily hours of nonagricultural work as a proportion of daily hours of work (agricultural hours of work, nonagricultural hours of work and household chores). Variable is only calculated for married households (i.e. with information for both head and spouse).
Hours of leisure 1	Hours spent by households resting, watching TV, or listening to the radio. Variable is only calculated for married households (i.e. with information for both head and spouse).
Hours of leisure 2	Hours spent by households resting, watching TV, listening to the radio, visiting neighbors, socializing, entertaining guests, and engaging in other leisure activities. Variable is only calculated for married households (i.e. with information for both head and spouse).

Electricity	
Expenditure on electricity as % of total expenditure	Expenditure on electricity as a percentage of total household expenditure.
Expenditure on electricity as % of total expenditure on energy	Energy sources include electricity, kerosene, candles, dry cell batteries, car batteries, liquefied petroleum gas, solar home system, firewood, animal dung and crop residues (hours of dung and residue collection were valued using urban/rural hourly wages for each department, reported in ENAHO 2004), electric generator set, charcoal, coal, and others.
Sources of energy	Number of sources of energy used by household from among 14 options: (1) electricity, (2) kerosene, (3) candles, (4) dry cell batteries, (5) car batteries, (6) liquefied petroleum gas, (7) solar home system, (8) firewood, (9) animal dung, (10) crop residues, (11) electric generator set, (12) charcoal, (13) coal, and (14) others.
Price	Soles/kW paid by household. Survey provides three options for reporting expenditure on electricity: Households paying a flat rate or by number of electrical appliances/light bulbs report their constant monthly payment. In the case of households paying per kW consumed, pollster requests their last three electricity bills. If respondents cannot show these bills, pollster asks for the approximate average payment of electricity during the 12 months previous to the survey. Price can only be estimated when respondents show their bills (which include their payment and the number of kW consumed).
Failures	Number of 30-minutes-or-more failures during the month prior to the survey.

Appendix B	Description of Variables (continued)
Variable	**Description**
Hours without electricity	Number of hours without electricity due to cuts or blackouts during the month prior to the survey.
Dimming in electricity service	Proportion of households reporting frequent dimming in electricity service.
Characteristics of town	
Access to phone	Information taken from the INEI's Pre-census 1999–2000. Public telephone available in town.
Type of road	Information taken from the INEI's Pre-census 1999–2000. Main type of road connecting the town with the capital of the district. Roads are classified in two categories: (1) *caminos carrozables, caminos de herradura, trocha,* rivers, and others; and (2) paved and *afirmada* roads
Secondary school	Information taken from the INEI's Pre-census 1999–2000. Secondary school available in town.
Quintile of population	Towns in the sample were classified in quintiles according to population reported in INEI's Pre-census 1999–2000.
Region	Coast, sierra, and jungle.
Characteristics of household	
Household size	Number of members of household.
Composition	Proportion of members 0–6, 7–14, 15–60, and 60 years or older.
Ethnicity	Self-reported ethnicity of household head among seven options: (1) native *quechua*, (2) native *aymara*, (3) native *amazonico*, (4) African-Peruvian; (5) Asian origin; (6) white; (7) *mestizo* (mixed).
Sex of head of household	Male or female head of household.
Education of head of household	Years of education of head of household.
Water supply in dwelling	Three possibilities for water supply used for drinking and cooking: (1) supply through public network; (2) well, pylon, or others; (3) river, springs, lakes, etc.

Appendix C — Balance of Variables in Common Support in Different Methods of Propensity Score Matching[a] (Nearest neighbor matching, caliper = 0.05)

	Unmatched sample		PS Match - household[b]		PS Match - household + town[c]		Two stage PSMatch[d]	
	Control	Treatment	Control	Treatment	Control	Treatment	Control	Treatment
Household characteristics								
Household size (members)	4.19	4.36**	4.33	4.33	4.45	4.31*	4.23	4.31
Proportion 6–13 years old	0.18	0.18	0.17	0.18	0.18	0.18	0.17	0.17
Proportion 14–60 years old	0.59	0.58	0.57	0.58	0.57	0.58	0.59	0.58
Proportion 60 years or older	0.14	0.16*	0.17	0.16	0.16	0.16	0.17	0.16
Indigenous HH head	0.45	0.19***	0.18	0.19*	0.20	0.19	0.18	0.18
Household with pipeline water	0.54	0.59***	0.58	0.59	0.60	0.59	0.64	0.67*
Other sources of water access	0.16	0.13**	0.15	0.13	0.11	0.14**	0.14	0.11**
Water from river, lake, etc	0.19	0.20	0.20	0.20	0.21	0.20	0.17	0.15
Years of education of HH head	7.07	6.97	6.94	6.99	6.98	7.00	7.11	7.35
Age of household head	48.91	50.91***	50.65	50.82	51.03	50.68	51.69	51.19
Female head of household	0.14	0.18***	0.17	0.18	0.16	0.18	0.15	0.16
Town characteristics								
Paved road in town	0.63	0.61	0.68	0.61***	0.67	0.62***		e
Public phone in town	0.54	0.46***	0.64	0.46***	0.49	0.47		e
Population (in thousands)	1.23	1.34**	1.42	1.33*	1.18	1.35***		e
Secondary school in town	0.56	0.52**	0.64	0.52***	0.52	0.52		e
Town on coast	0.39	0.59***	0.51	0.59***	0.61	0.59		e
Town in highlands	0.47	0.36***	0.32	0.36***	0.34	0.37		e
Town in jungle	0.14	0.05***	0.17	0.05***	0.05	0.05		e
Observations	1188	1483	1476	1476***	1437	1437	906	906

[a] For detailed definition of variables, please see Appendix B.
[b] Probit regression includes household variables.
[c] Probit regression includes household and town variables.
[d] First stage: towns with same characteristics (roads, public phone, quintile of population, secondary school, and region). Second stage: propensity score matching within households in comparable towns.
[e] By definition, town variables are the same in the two-step matching.

* Significant at a 10% level of confidence; ** Significant at a 5% level of confidence; *** Significant at a 1% level of confidence.

Appendix D — Summary Statistics of Performance, Quality of Service and Welfare Indicators (unmatched)

| | Unmatched sample | | | | |
| | Treatment | | Control | | |
	Mean	Obs	Mean	Obs	Diff[1]
Expenditure on electricity (% of expenditure on energy)	0.48	1473	0.47	1186	0.01
Expenditure on electricity (% of total expenditure)	0.05	1483	0.05	1188	0.00
Number of sources of electricity	2.83	1483	3.09	1188	−0.26**
Price per kW	1.00	499	1.08	293	−0.08
Number of monthly failures (30+ min)	1.30	1408	1.32	1073	−0.02
Often dimming in electric service	0.11	1483	0.13	1188	−0.03**
Monthly hours of blackouts	4.59	1312	5.62	1028	−1.03**
Hours of work – HH	10.90	1061	12.54	876	−1.64**
Hours of work – HH (incl chores)	17.27	1061	19.16	876	−1.89**
Proportion of non-farm hours of work – HH	0.49	1012	0.39	859	0.09**
Proportion of non-farm hours of work – HH (incl chores)	0.29	1057	0.25	876	0.04**
Hours of work – indiv (incl chores)	5.45	2168	6.26	1799	−0.81**
Hours of work – indiv	8.63	2168	9.59	1799	−0.96**
Proportion of non-farm hours of work – individual	0.50	1727	0.42	1605	0.08**
Proportion of non-farm hours of work – individual (incl chores)	0.27	2097	0.25	1775	0.02**
Per-capita expenditure	217.25	1447	198.24	1147	19.01**
Proportion of non-farm income	0.40	906	0.33	823	0.07**
Hours of leisure (TV and radio)	4.42	876	4.17	1061	0.24**
Hours of leisure (TV, radio, socializing, and others)	9.16	876	7.71	1061	1.46**

[1] The table presents the unmatched means of interest variables in Tables 11, 12 and 13.
** denotes significance at a 95% level of confidence.
For variable definitions, please see Appendix B.

Bibliography

Acosta, O.L. and I. Fainboim, editors. 1994. *Las Reformas Económicas del Gobierno del Presidente Gaviria: Una Visión Desde Adentro*. Bogotá, Colombia: Ministerio de Hacienda y Crédito Público.

Adam, C., W. Cavendish and P. Mistry. 1992. *Adjusting Privatization: Case Studies From Developing Countries*. London, United Kingdom and Portsmouth, United States: J. Curry and Heinemann.

Aguas Argentinas. 2003. "Hacia un acceso universal a los servicios. El modelo participativo de gestión." Buenos Aires, Argentina: Aguas Argentinas. Mimeographed document.

Aguilar, G. 2003. "El sistema tarifario del servicio público de electricidad: Una evaluación desde el punto de vista de los usuarios." Documento de Trabajo 224. Lima, Peru: Pontificia Universidad Católica del Perú.

Aharoni, Y. 1990. "On the Measurement of Successful Privatization." In: R. Ramamurti and R. Vernon, editors. *Privatization and Control of State-Owned Enterprises*. EDI Development Studies. Washington, DC: World Bank.

Alcázar, L. 2004. *Evaluation of the Concessions and Privatization Processes in Infrastructure Services in Peru*. Lima, Peru: Grupo de Análisis para el Desarrollo (GRADE).

Alcázar, L., M. Abdala and M. Shirley. 2002. "The Buenos Aires Water Concession." In: M. Shirley, editor. *Thirsting for Efficiency: The Economics and Politics of Urban Water System Reform*. Amsterdam, The Netherlands: Pergamon.

Alcázar, L., E. Nakasone and M. Torero. 2007. "Provision of Public Services and Welfare of the Poor: Learning from an Incomplete Electricity Privatization Experience in Rural Peru." Research Network Working Paper R-526. Washington, DC: Inter-American Development Bank.

Allen, B. 1990. "Information as an Economic Commodity." *American Economic Review* 80(2): 268–273.

Almond, D., K. Chay and D. Lee. 2005. "The Costs of Low Birth Weight." *Quarterly Journal of Economics* 120(3): 1031–1083.

Altimir, O. and L. Beccaria. 1998. "Effects of Macroeconomic Changes and Reforms on Urban Poverty in Argentina." In: E. Ganuza, L. Taylor and S. Morley, editors. *Macroeconomic Policy and Poverty in Latin America and the Caribbean*. Madrid: Grupo Mundi-Prensa.

Andrés, L., J. Guasch and V. Foster. 2004. "The Impact of Privatization on Firms in the Infrastructure Sector in Latin-American Countries." Washington, DC: World Bank. Mimeographed document.

Andrew, T.N. and D. Petkov. 2003. "The Need for a Systems Thinking Approach to the Planning of Rural Telecommunications Infrastructure." *Telecommunications Policy* 27: 75–93.

Angeles, G., A.J. Trujillo and A. Lastra. 2007. "Is Public Health Expenditure in Ecuador Progressive or Regressive?" *International Journal of Public Policy* 2(3/4): 186–216.

Artana, D., F. Navajas and S. Urbiztondo. 2000. "Governance and Regulation in Argentina." In: W. Savedoff and P. Spiller, editors. *Spilled Water*. Washington, DC: Inter-American Development Bank.

Asociación Nacional de Instituciones Financieras (ANIF) and Centro Latinoamericano para la Privatización. 1992. "Privatización y Re-privatización en Colombia, Teoría y Práctica." Cali, Colombia: ANIF.

Babcock, B.A. 1990. "The Value of Weather Information in Market Equilibrium." *American Journal of Agricultural Economics* 72(1): 63–72.

Barberis, N. et al. 1996. "How Does Privatization Work? Evidence from the Russian Shops." *Journal of Political Economy* 104: 769–790.

Barker, D. 1998. *Mothers, Babies and Health in Later Life*. Edinburgh, UK: Churchill Livingstone.

Barrera-Osorio, F. and M. Olivera. 2007. "Does Society Win or Lose as a Result of Privatization? Provision of Public Services and Welfare of the Poor: The Case of Water Sector Privatization." Research Network Working Paper R-525. Washington, DC: Inter-American Development Bank.

Bayes, A. 2001. "Infrastructure and Rural Development: Insights from a Grameen Bank Village Phone Initiative in Bangladesh." *Agricultural Economics* 25 (2–3): 261–272.

Bayliss, K. 2001. "Privatization of Electricity Distribution: Some Economic, Social and Political Perspectives." Public Service International Research Unit Report. Greenwich, United Kingdom, University of Greenwich, Policy Studies Institute.

Behrman, J. and B. Wolfe. 1987. "How Does Mother's Schooling Affect Family Health, Nutrition, Medical Care Usage and Household Sanitation?" *Journal of Econometrics* 36(1–2): 185–204.

Bel, G. and M. Warner. 2006. "Local Privatization and Costs: A Review of Empirical Evidence." Paper presented at the "Privatization and Local Government Reform" International Workshop, Barcelona, Spain, June.

Benitez, D., O. Chisari and A. Estache. 2003. "Can the Gains from Argentina's Utilities Reform Offset Credit Shocks?" In: C. Ugaz and C. Waddams Price, editors. *Utility Privatisation and Regulation—A Fair Deal for Consumers?* Aldershot, United Kingdom: Edward Elgar.

Bertrand, M., E. Duflo and S. Mullainathan. 2004. "How Much Should We Trust Differences-in-Differences Estimates." *Quarterly Journal of Economics* 119 (1): 249–275.

Bertrand, M. and S. Mullainathan. 2004. "Are Emily and Greg More Employable than Lakisha and Jamal? A Field Experiment on Labor Market Discrimination." *American Economic Review* 94(4): 991–1013.

Binswanger, H., S. Khandker and M. Rosenzweig. 1993. "How Infrastructure and Financial Institutions Affect Agricultural Output and Investment in India." *Journal of Development Economics* 41: 337–366.

Birch, M. and J. Haar, editors. 2000. *The Impact of Privatization in the Americas*. Coral Gables, FL: University of Miami, North South Center Press.

Bird, R. and M. Smart. 2002. "Intergovernmental Fiscal Transfers: International Lessons for Developing Countries." *World Development* 30(6): 899–912.

Birdsall, N. and J. Nellis. 2002. "Winners and Losers: Assessing the Distributional Impact of Privatization." Working Paper 6. Washington, DC: Center for Global Development.

Birdsall, N., and J. Nellis, editors. 2005. *Reality Check: The Distributional Impact of Privatization in Developing Countries*. Washington, DC: Center for Global Development.

Blank, R.M. 1999. "When Can Public Policy Makers Rely on Private Markets? The Effective Provision of Social Security." NBER Working Paper 7099. Cambridge, MA: National Bureau of Economic Research.

Boardman, A. and A. Vining. 1989. "Ownership and Performance in Competitive Environments: A Comparison of the Performance of Private, Mixed, and State-Owned Enterprises." *Journal of Law and Economics* 32: 1–33.

Bonifaz, J. 2001. "Distribución eléctrica en el Perú: Regulación y eficiencia." Lima Peru: Consorcio de Investigación Económica y Social (CIES)/Centro de Investigación de la Universidad del Pacífico (CIUP).

Borcherding, T., W. Pommerehene and F. Schneider. 1982. "Comparing the Efficiency of Private and Public Production: The Evidence from Five Countries." *Zeitschrift für Nationalokonomie* 2: 127–156.

Bortolotti, B., M. Fantini and D. Siniscalco. 2000. "Privatisations and Institutions: A Cross-Country Analysis." CESifo Working Paper 375. Munich, Germany: Center for Economic Research.

Botton, S., A. Braïlowsky and S. Matthieussent. 2003. "Los verdaderos obstáculos para el acceso universal al servicio de agua en los países emergentes." Buenos Aires, Argentina: Aguas Argentinas. Mimeographed document.

Bouille, D., H. Dubrovsky and C. Maurer. 2002. "Argentina: Market Driven Reform of the Electricity Sector." In: N. Navroz, editor. *WRI Power Politics*. Washington, DC: World Resources Institute.

Boycko, M., A. Shleifer and R. Vishny. 1993. *A Theory of Privatization*. Cambridge, MA: MIT Press.

Boyne, G. 1998. *Public Choice Theory and Local Government: A Comparative Analysis of the UK and the USA*. New York: St. Martin's Press.

Bravo, D., C. Sanhueza and S. Urzúa. 2006. "Is There Labor Market Discrimination among Professionals in Chile? Lawyers, Doctors and Business-People." Santiago, Chile: Universidad de Chile, Departamento de Economía. Manuscript.

Brown, J.D., J. Earle and A. Telegdy. 2005. "The Productivity Effects of Privatization: Longitudinal Estimates from Hungary, Romania, Russia, and Ukraine." Upjohn Institute Working Paper 05–121. Kalamazoo, MI: Upjohn Institute for Employment Research.

Brown, J.D., J. Earle and V. Vakhitov. 2006. "Wages, Layoffs, and Privatization: Evidence from Ukraine." CERT Discussion Papers 601. Edinburgh, United Kingdom: Heriot-Watt University, Center for Economic Reform and Transformation.

Caballero, Rafael. 1999. "Balance de la Reforma del Sector Eléctrico, las Privatizaciones y el Marco Regulatorio, el Caso Colombiano." Bogotá, Colombia: Departamento Nacional de Planeación.

Calvó-Armengol, A. and M.O. Jackson. 2004. "The Effects of Social Networks on Employment." *American Economic Review* 94(3): 426–454.

Caplan, K., B. Evans and J. McMahon. 2004. "The Partnership Paper Chase: Structuring Partnership Agreements in Water and Sanitation in Low-Income Communities." London, United Kingdom: Building Partnerships for Development in Water and Sanitation.

Cárdenas, J.-C. et al. 2006. "Discrimination in the Provision of Social Services to the Poor: A Field Experimental Study." Bogotá, Colombia: Universidad de los Andes, Centro de Estudios sobre Desarrollo Económico (CEDE). Manuscript.

Carrera, J., D. Checchi and M. Florio. 2004. "Privatization Discontent and its Determinants, Evidence from Latin America." GIUGNO Working Paper 23.2004. Milan, Italy: Università degli Studi di Milano, Departamento di Economia Politica y Aziendale,.

Carrillo, P., O. Bellettini and E. Coombs. "Stay Public or Go Private? A Comparative Analysis of Water Services between Quito and Ecuador." Research Network Working Paper R-538. Washington, DC: Inter-American Development Bank.

Castillo, M., R. Petrie and M. Torero. 2007. "Ethnic and Social Barriers to Cooperation: Experiments Studying the Extent and Nature of Discrimination in Urban Peru." Atlanta, GA and Lima, Peru: Georgia Institute of Technology, Georgia State University, International Food Policy Research Institute (IFPRI), and Grupo de Análisis para el Desarrollo (GRADE). Manuscript.

Chaparro, J.C., M. Smart and J.G. Zapata. 2006. "Intergovernmental Transfers and Municipal Finance in Colombia." In: R. Bird, J. Poterba and J. Slemrod, editors. *Fiscal Reform in Colombia*. Cambridge, MA: MIT Press.

Chisari, O., A. Estache, and C. Romero. 1999. "Winners and Losers from Utility Privatization in Argentina: Lessons from a General Equilibrium Model." *The World Bank Economic Review* 13(2): 357–378.

Chong, A., V. Galdo and M. Torero. 2006. "Access to Telephone Services and Household Income in Poor Rural Areas Using a Quasi-Natural Experiment for Peru." Research Department Working Paper 535. Washington, DC: Inter-American Development Bank.

Chong, A., J. Hentschel and J. Saavedra. 2004. "Bundling in the Provision of Public Services: The Case of Peru." Washington, DC: World Bank. Manuscript.

Chong, A. and G. León. 2007. "Privatized Firms, Rule of Law, and Labor Outcomes in Emerging Markets." Research Department Working Paper 608. Washington, DC: Inter-American Development Bank.

Chong, A. and F. López-de-Silanes. 2003. "Privatization and Labor Restructuring around the World." New Haven, CT: Yale University, School of Management. Manuscript.

———. 2004. "Privatization in Latin America: What Does the Evidence Say?" *Economía* 4(2): 37–111.

Chong, A. and F. López-de-Silanes, editors. 2005. *Privatization in Latin America: Myths and Reality*. Stanford, CA: Stanford University Press.

Chong, A., F. López-de-Silanes and M. Torero. 2007. "Back to Reality: What Happens with Workers After Privatization?" Washington, DC: Inter-American Development Bank. Manuscript.

Chong, A., E. Nakasone and M. Torero. 2007. "Provision of Public Services and Welfare of the Poor: Learning from an Incomplete Electricity

Privatization Process in Rural Peru." Research Department Working Paper 526. Washington, DC: Inter-American Development Bank.

Chong, A., V. Rodriguez and M. Torero. 2007. "Biased Retrenchment and Post-Privatization Wages in Mexico." Washington, DC: Inter-American Development Bank. Manuscript.

Clarke, G., K. Kosec and S. Wallsten. 2004. "Has Private Participation in Water and Sewerage Improved Coverage? Empirical Evidence from Latin America." World Bank Policy Research Working Paper 3445. Washington, DC: World Bank.

Comisión de Control Cívico de la Corrupción. 2005. Informe de Investigación: "Presuntas irregularidades en el sistema de tratamiento y abastecimiento de agua potable distribuida en el Suburbio Sur-Oeste de la ciudad de Guayaquil por parte de Interagua, concesionaria de ECAPAG, como causa coadyuvante del brote de Hepatitis A, aparecido en el Suburbio Oeste de la ciudad de Guayaquil." September 29. Guayaquil, Ecuador: Comisión de Control Cívico de la Corrupción.

Comisión Nacional para la Privatización (COPRI). 2000. Evaluación del proceso de privatización del sector eléctrico. Lima, Peru: COPRI.

Consejo Superior de Política Fiscal. 2002. "Privatizaciones y Concesiones de la Nación." Bogotá, Colombia: Ministerio de Hacienda y Crédito Público.

Constance, P. 2003. "A Fair Price." IDB America: Magazine of the Inter-American Development Bank (November).

Cotlear, D. 1989. Desarrollo Campesino en los Andes. Lima, Peru: Instituto de Estudios Peruanos.

Davidson, R. and J. MacKinnon. 1999. "Bootstrap Testing in Nonlinear Models." International Economic Review 40(2): 487–508.

De Ferranti, D. et al. 2005. Beyond the City: The Rural Contribution to Development. World Bank Latin American and Caribbean Studies. Washington, DC: World Bank.

Defensoría del Pueblo, Ecuador. 2005. Defensoría Adjunta Segunda del Litoral y Galápagos. Expediente Defensorial No. 314-DASLG-05. Guayaquil, September 28.

Delfino, J. and A. Casarin. 2003. "The Reform of the Utilities Sector in Argentina." In: C. Ugaz and C. Waddams Price, editors. Utility

Privatisation and Regulation: A Fair Deal for Consumers? Aldershot, United Kingdom: Edward Elgar.

Departamento Nacional de Planeación. 1996. "Conpes 2885: Mecanismos para la Promoción de las Privatizaciones." Bogotá, Colombia: Departamento Nacional de Planeación.

———. 1997. "Conpes 2929: Balance de los Procesos de Vinculación de Capital Privado—Las Privatizaciones." Bogotá, Colombia: Departamento Nacional de Planeación.

———. 1998. "Conpes 3006: Participación Privada en Agua Potable: Seguimiento." Bogotá, Colombia: Departamento Nacional de Planeación.

———. 2005. "Inversión Privada en Infraestructura 1993–2003." Bogotá, Colombia: Dirección de Infraestructura y Energía Sostenible, Proyecto Gerencia de Participación Privada en Infraestructura.

Donahue, J. 1989. *The Privatization Decision: Public Ends, Private Means*. New York: Basic Books.

D'Souza, J. and W. Megginson. 1999. "The Financial and Operating Performance of Privatized Firms During the 1990s." *Journal of Finance* 54(4): 1397–1468.

Durlauf, S.N. 2002. "On the Empirics of Social Capital." *Economic Journal* 112(483): 459–479.

Earle, J. and S. Gelbach. 2003. "A Spoonful of Sugar: Privatization and Popular Support for Reform in the Czech Republic." *Economics and Politics* 15(1): 1–32.

Empresa de Administración de Infraestructura Eléctrica. (ADINELSA). 2004. *Memoria Anual*. Lima, Peru: ADINELSA.

Empresa Cantonal de Agua Potable y Alcantarillado de Guayaquil (ECAPAG). Undated. "Informe de Gestión 2001–2005." Guayaquil, Ecuador: ECAPAG.

———. Undated. "Proceso de Modernizacion de los Servicios Publicos de Agua Potable y Saneamiento de Guayaquil al Sector Privado: 1994–2001." Guayaquil, Ecuador: ECAPAG.

———. Undated. *Unidad Ejecutoria del Proyecto* (IDB). "Informe Semestral del Programa de Concesión al Sector Privado de los Servicios de Agua Potable y Alcantarillado de Guayaquil, Financiado con el Préstamo

BID 1026/OC-EC. Período de Julio-Diciembre del 2003. Guayaquil, Ecuador: ECAPAG.

Económica Consultores. 2004. "Diagnóstico del impacto sobre cada sector (acueducto, alcantarillado, aseo, energía, gas y telecomunicaciones), de la legislación y normatividad específica que en materia de servicios públicos domiciliarios se expidió en cumplimiento de los preceptos constitucionales de 1991." Informe final. Bogotá, Colombia: Económica Consultores.

Elías, J., V. Elías and L. Ronconi. 2007. "Determinants of Popularity among Adolescents in Argentina." Research Network Working Paper R-539. Washington, DC: Inter-American Development Bank.

Engel, E., R. Fischer and A. Galetovic. 2003. "Privatizing Highways in Latin America: Fixing What Went Wrong." *Economía* 4(1): 129–164.

Escobal, J. 2002. "The Determinants of Nonfarm Income Diversification in Rural Peru." *World Development* 29(3): 497–508.

Esrey, S., and R. Feachem. 1989. "Interventions for the Control of Diarrhoeal Bases among Young Children: Promotion of Food Hygiene." WHO/CDD/89.30. Geneva: World Health Organization.

Esrey, S. et al. 1991. "Effects of Improved Water Supply and Sanitation on Ascariasis, Diarrhea, Dracunculiasis, Hookworm Infection, Schistosomiasis and Trachoma." *Bulletin of the World Health Organization* 69(5): 609–621.

Fainboim, I. 2000. "Efectos sobre el bienestar social de la reestructuración y capitalización de la Empresa de Energía de Bogotá." Bogota, Colombia: Fedesarrollo. Manuscript.

Feather, P.M. and G.S. Amacher. 1994. "Role of Information in the Adoption of Best Management Practices for Water Quality Improvement." *Agricultural Economics* 11(2–3): 159–170.

Fernández, D. 2004. "Sector agua potable en Colombia: Desarrollo económico reciente en infraestructura." Finance, Private Sector and Infrastructure Unit Report. Washington, DC: World Bank.

Fiszbein, A. 1997. "The Emergence of Local Capacity: Lessons From Colombia." *World Development* 25(7): 1029–1043.

Forero, J. 2005. "Latin America Fails to Deliver on Basic Needs." *New York Times*, February 22, A1.

Foster, V. 2002. "Ten Years of Water Service Reform in Latin America: Towards an Anglo-French Model." In: P. Seidstat, D. Haarmeyer, and S. Hakin, editors. *Reinventing Water and Wastewater Systems: Global Lessons for Improving Management*. New York: John Wiley & Sons, Inc.

———. 2005. "Ten Years of Water Service Reform in Latin America: Toward an Anglo-French Model." Water and Sanitation Sector Board Discussion Paper Series, Paper 3. Washington, DC: World Bank.

Freije, S. and L. Rivas. 2003. "Privatization, Inequality and Welfare: Evidence from Nicaragua." IPD Working Paper 2003–03. Puebla, Mexico: Universidad de las Américas.

Frydman, R. et al. 1999. "When Does Privatization Work? The Impact of Private Ownership on Corporate Performance in Transition Economies." *Quarterly Journal of Economics* 114(4): 1153–1191.

Galal, A. et al. 1994. *Welfare Consequences of Selling Public Enterprises: An Empirical Analysis*. Washington, DC: Oxford University Press/World Bank.

Galiani, S., P. Gertler and E. Schargrodsky. 2005. "Water for Life: The Impact of the Privatization of Water Services on Child Mortality." *Journal of Political Economy* 113(1): 83–120.

Galiani, S. et al. 2005. "The Benefits and Costs of Privatization in Argentina: A Microeconomic Analysis." In: A. Chong and F. López-de-Silanes, editors. *Privatization in Latin America: Myths and Realities*. Stanford, CA, and Washington, DC: Stanford University Press and World Bank.

Galiani, S., M. González-Rozada and E. Schargrodsky. 2007. "Water Expansions in Shantytowns: Health and Savings." Research Network Working Paper R-527. Washington, DC: Inter-American Development Bank.

Galiani, S. and F. Sturzenegger. 2005. "The Impact of Privatization on the Earnings of Restructured Workers." Documento de Trabajo 05/2006. Buenos Aires, Argentina: Universidad Torcuato di Tella, Escuela de Negocios.

Gallardo, J. and L. Bendezú. 2005. "Evaluación del Fondo Social de compensación eléctrica." Documento de Trabajo 7. Lima, Peru: Organismo Superior de la Inversión en Energía (OSINERG).

Gandelman, E., N. Gandelman and J. Rothschild. 2007. Gender Differentials in Judicial Proceedings: Field Evidence from Housing-Related Cases in Uruguay. Working Paper. Universidad ORT Uruguay. Montevideo.

Gaviria, A. 2006. "Movilidad social y preferencias por redistribución en América Latina." Documento CEDE 2006–03. Bogotá, Colombia: Universidad de los Andes.

Gerchunoff, P., E. Greco and D. Bondorevsky. 2003. "Comienzos diversos, distintas trayectorias y final abierto: más de una década de privatizaciones en Argentina, 1990–2002." Serie Gestión Pública 34. Santiago, Chile: United Nations Economic Commission for Latin America and the Caribbean.

Godtland, E. et al. 2004. "The Impact of Farmer Field Schools on Knowledge and Productivity: A Study of Potato Farmers in the Peruvian Andes." *Economic Development and Cultural Change* 53(1): 63–92.

Gómez-Lobo, A. and D. Contreras. 2003. "Water Subsidy Policies: A Comparison of the Chilean and Colombian Schemes. " *World Bank Economic Review* 17(3): 391–407.

Gómez-Lobo, A. and M. Melendez. 2007. "Social Policy, Regulation and Private Sector Water Supply: The Case of Colombia." Serie Documentos de Trabajo 252. Santiago, Chile: Universidad de Chile, Departamento de Economía.

González-Eiras, M. and M. Rossi. 2007. "The Impact of Electricity Sector Privatization on Public Health." Research Network Working Paper R-524. Washington, DC: Inter-American Development Bank.

Gray, P. 1997. "Colombia's Gradualist Approach to Private Participation in Infrastructure." Private Sector Development Department Note 113. Washington, DC: World Bank.

Grupo Sophia. 1999. "Acerca de las posibilidades de control ciudadano de los servicios públicos privatizados." Chapter 3 in *Mecanismos de Social Control*. Buenos Aires, Argentina: Grupo Sophia.

Guasch, J. and P. Spiller. 1999. *Managing the Regulatory Process: Design, Concepts, Issues, and the Latin America and Caribbean Story*. World Bank Latin American and Caribbean Studies. Washington, DC: World Bank.

Hachette, D. and R. Luders. 1992. *La Privatización en Chile*. Oakland, CA: Institute for Contemporary Studies Press/Centro Internacional para el Desarrollo Económico.

Haskel, J. and S. Szymanski. 1992. "A Bargaining Theory of Privatisation." *Annals of Public and Cooperative Economics* 63: 207–227.

Heckman, J. and J. Hotz. 1989. "Choosing Among Alternative Nonexperimental Methods for Estimating the Impact of Social Programs: The Case of Manpower Training." *Journal of the American Statistical Association* 84: 862–880.

Heckman, J., H. Ichimira and P. Todd. 1997. "Matching as an Econometric Evaluation Estimator: Evidence from Evaluating a Job Training Program." *Review of Economic Studies* 64: 605–654.

———. 1998. "Matching as an Econometric Evaluation Estimator." *Review of Economic Studies* 65(2): 261–94.

Heckman, J., H. Ichimira, J. Smith and P. Todd. 1996. "Characterizing Selection Bias Using Experimental Data." *Econometrica* 66: 1017–1098.

Heckman, J., R. Lalonde and J. Smith. 1999. "The Economics and Econometrics of Active Labor Market Programs." In: O. Ashenfelter and D. Card, editors. *Handbook of Labor Economics*. Volume 3a. New York: Elsevier Science.

Heckman, J. and J. Smith. 1995. "Assessing the Case of Social Experiments." *Journal of Economic Perspectives* 9: 85–110.

Hefetz, A. and M. Warner. 2004. "Privatization and Its Reverse: Explaining the Dynamics of the Government Contracting Process." *Journal of Public Administration Research and Theory* 14(2): 171–190.

Hodge, G.A. 2000. *Privatization: An International Review of Performance*. Boulder, CO: Westview Press.

Howe, J. and P. Richards. 1984. *Rural Roads and Poverty Alleviation: A Study Prepared for the International Labour Office within the Framework of the World Employment Programme*. London: Intermediate Technology Publications.

Hubbard, T.N. 2003. "Information, Decisions, and Productivity: On-Board Computers and Capacity Utilization in Trucking." *American Economic Review* 93(4):1328–1353.

Huffman, W. and S. Mercier. 1991. "Adoption of Microcomputers Technologies: An Analysis of Farmers' Decisions." *Review of Economic and Statistics* 73(3): 541–546.

Hujer, R. and M. Caliendo. 2000. "Evaluation of Active Labour Market Policy: Methodological Concepts and Empirical Estimates." IZA Discussion Paper 236. Bonn, Germany: Institute for the Study of Labor (IZA).

Hujer, R. and M. Wellner. 2000. "The Effects of Public Sector Sponsored Training on Individual Employment Performance in East Germany." IZA Discussion Paper 141. Bonn, Germany: Institute for the Study of Labor (IZA).

Instituto Nacional de Estadistica y Censos (INDEC). 2003, 2004. "Incidencia de la pobreza y de la indigencia en los aglomerados urbanos." Press Releases. Buenos Aires, Argentina: INDEC.

Instituto Nacional de Estadisticas y Censos de Ecuador (INEC). Various dates. Surveys used from August 1994–August 1995, 2001 and 2004. Quito, Ecuador: INEC.

Instituto Peruano de Economía. 2004. *La infraestructura que necesita el Perú: Brecha de inversión en infraestructura de servicios públicos*. Lima, Peru: Instituto Peruano de Economía.

Inter-American Development Bank (IDB). 2002. "The Privatization Paradox." *Latin American Economic Policies* 18. Washington, DC: IDB.

———. 2007. *Outsiders? The Changing Pattern of Exclusion in Latin American and the Caribbean*. Report on Economic and Social Progress in Latin America. Washington, DC: IDB.

International Telecommunications Union (ITU). 1998. *World Telecommunication Development Report: Universal Access*. Geneva, Switzerland: ITU.

———. 2003. *World Telecommunication Development Report: Access Indicators for the Information Society*. Geneva, Switzerland: ITU.

Isham, J. 2002. "The Effect of Social Capital on Technology Adoption: Evidence from Rural Tanzania." *Journal of African Economies* 11(1): 39–60.

Jacoby, H. 1998. "Access to Markets and the Benefits of Rural Roads." Washington, DC: World Bank Development Research Group Rural Development.

———. 2000. "Access to Markets and the Benefits of Rural Roads." *Economic Journal* 110(465): 713–737.

Jalan, J. and M. Ravallion. 2003. "Does Piped Water Reduce Diarrhea for Children in Rural India?" *Journal of Econometrics* 112(1): 153–173.

Jiménez, E. 1995. "Human and Physical Infrastructure: Public Investment and Pricing Policies in Developing Countries." In: J. Behrman and T.N. Srinivasan, editors. *Handbook of Development Economics*. Volume 3. Amsterdam, The Netherlands: North-Holland.

Kahneman, D. and A. Tversky. 1979. "Prospect Theory: An Analysis of Decisions under Risk." *Econometrica* 47: 313–327.

Kebede, Y., K. Gunjal and G. Coffin. 1990. "Adoption of New Technologies in Ethiopian Agriculture: The Case of Tegulet-Bulga District Shoa Province." *Agricultural Economics* 4(1): 27–43.

Kikeri, S. 1999. "Privatization and Labor: What Happens to Workers When Government Divests?" World Bank Technical Paper 396. Washington, DC: World Bank.

Kikeri, S., J. Nellis and M. Shirley. 1994. "Privatization: The Lessons from Market Economies." *World Bank Research Observer* 9: 241–272.

La Porta, R. and F. López-de-Silanes. 1999. "The Benefits of Privatization: Evidence from Mexico." *Quarterly Journal of Economics* 114(4): 1193–1242.

Latinobarómetro. Various years. Available at:
http://www.latinobarometro.org

Lavy, V. et al. 1996. "Quality of Health Care, Survival and Health Outcomes in Ghana." *Journal of Health Economics* 15(3): 333–357.

Lebo, J. and D. Schelling. 2001. "Design and Appraisal of Rural Transport Infrastructure: Ensuring Basic Access for Rural Communities." World Bank Technical Paper 496. Washington, DC: World Bank.

Lechner, M. 2000. "Programme Heterogeneity and Propensity Score Matching: An Application to the Evaluation of Active Labour Market Policies." Econometric Society World Congress Contributed Papers 0647. Evanston, IL: Econometric Society.

Lee, L.-F., M. Rosenzweig and M. Pitt. 1997. "The Effects of Improved Nutrition, Sanitation, and Water Quality on Child Health in High-Mortality Populations." *Journal of Econometrics* 77(1): 209–235.

Leff, N.H. 1984. "Externalities, Information Costs, and Social Benefit-Cost Analysis for Economic Development: An Example from Telecomm." *Economic Development and Cultural Change* 32: 255–276.

Leipziger, D. 2004. "The Privatization Debate: The Case of Latin American Utilities." Paper presented at the Joint Study on Infrastructure Development in East Asia, First Regional Workshop, Manila, Philippines. January 15–16.

Lipton, M. and M. Ravallion. 1995. "Poverty and Policy." In: J. Behrman and T.N. Srinivasan, editors. *Handbook of Development Economics*. Volume 3. Amsterdam, The Netherlands: North-Holland.

López-de-Silanes, F. 1997. "Determinants of Privatization Prices." *Quarterly Journal of Economics* 112(4): 965–1025.

López-de-Silanes, F. and G. Zamarripa. 1995. "Deregulation and Privatization of Commercial Banking: Pre versus Post-Performance." *Review of Economic Analysis/Revista de Análisis Económico* 10(2): 113–164.

Lora, E. 2001. "Structural Reforms in Latin America: What Has Been Reformed and How to Measure It." Research Department Working Paper 466. Washington, DC: Inter-American Development Bank.

Lora, E., editor. 2007. *The State of State Reform in Latin America*. Stanford, CA and Washington, DC: Stanford University Press and Inter-American Development Bank.

Lora, E. and U. Panizza. 2002. "Structural Reforms in Latin America Under Scrutiny." Research Department Working Paper 470. Washington, DC: Inter-American Development Bank.

Lowery, D. 1998. "Consumer Sovereignty and Quasi-Market Failure." *Journal of Public Administration Research and Theory* 8(2):137–172.

Luders, R. 1991. "Chile's Massive SOEs Divestiture Program, 1975–1990: Failures and Successes." *Contemporary Policy Issues* 9(4): 1–19.

Mackenzie, G. 1998. "The Macroeconomic Impact of Privatization." *IMF Staff Papers* 45(2): 363–373.

Mardones-Santander, F. et al. 1988. "Effect of a Milk-Based Food Supplement on Maternal Nutritional Status and Fetal Growth in Underweight Chilean Women." *American Journal of Clinical Nutrition* 47(3): 413–419.

Matambalya, F. and S. Wolf. 2001. "The Role of ICT for the Performance of SMEs in East Africa: Empirical Evidence from Kenya and Tanzania." ZEF Discussion Papers on Development Policy 42. Bonn, Germany: Center for Development Research (ZEF).

Mayer, E. 2002. *The Articulate Peasant: Household Economies in the Andes*. Boulder, CO: Westview Press.

McKenzie, D. and D. Mookherjee. 2003. "The Distribution Impact of Privatization in Latin America: Evidence from Four Countries." *Economía* 3(2): 161–233.

Megginson, W., R. Nash and M. van Randenborgh. 1994. "The Financial and Operating Performance of Newly Privatized Firms: An International Empirical Analysis." *Journal of Finance* 49(2): 403–452.

Megginson, W. and J. Netter. 2001. "From State to Market: A Survey of Empirical Studies on Privatization." *Journal of Economic Literature* 39: 321–389.

Merrick, T. 1985. "The Effect of Piped Water on Early Childhood Mortality in Urban Brazil, 1970 to 1976." *Demography* 22(1): 1–24.

Millan, J., E. Lora and A. Micco. 2001. "Reforms in Latin America Sustainability of the Electricity Sector." Paper prepared for the seminar "Towards Competitiveness: The Institutional Path," Santiago, Chile.

Ministerio de Energía y Minas. Various dates *Anuario estadístico*. Lima, Peru: Ministerio de Energía y Minas.

Ministerio de Energía y Minas, Dirección Ejecutiva de Proyectos. 2004. *Plan Nacional de Electrificación Rural 2005–2014*. Lima, Peru: Ministerio de Energía y Minas.

Moreno, M., H. Ñopo, J. Saavedraa and M. Torero. 2004. "Gender and Racial Discrimination in Hiring: A Pseudo Audit Study for Three Selected Occupations in Metropolitan Lima." IZA Discussion Paper 979. Bonn, Germany: Institute for the Study of Labor (IZA).

Moulton, B. 1990. "An Illustration of a Pitfall in Estimating the Effects of Aggregate Variables in Micro Units." *Review of Economics and Statistics* 72(2): 334–338.

Nelson, M. 1999. "Nutrition and Health Inequalities." In: D. Gordon et al., editors. *Inequalities in Health: The Evidence Presented to the Independent*

Inquiry into Inequalities in Health. Studies in Poverty, Inequality and Social Exclusion Series. Bristol, United Kingdom: Policy Press.

Neu, D. and A.S. Rahaman. 2003. "Accounting, Dependency, International Financial Institutions, and Privatization of Water Services in a Developing Country." Paper prepared for the 2003 Interdisciplinary Perspectives on Accounting conference, Madrid, Spain, July 13–16.

Nickson, A. 2001. "Establishing and Implementing a Joint Venture: Water and Sanitation Services in Cartagena, Colombia." Working Paper 442 03. London, United Kingdom: GHK International and DFID.

Niskanen, W.A. 1971. *Bureaucracy and Representative Government*. Chicago: Aldine.

Noll, R., M. Shirley and S. Cowan. 2000. "Reforming Urban Water Systems in Developing Countries." In: A. Krueger, editor. *Economic Policy Reform: The Second Stage*. Chicago: University of Chicago Press.

Ñopo, H., J. Saavedra and M. Torero. 2007. "Ethnicity and Earnings in a Mixed-Race Labor Market." *Economic Development and Cultural Change* 55: 709–734.

Núñez, J. and R. Gutiérrez. 2004. "Classism, Dscrimination and Meritocracy in the Labor Market: The Case of Chile." Working Paper 208. Santiago, Chile, Universidad de Chile, Facultad de Ciencias Económicas y Administrativas, Departamento de Economía.

Observatorio Ciudadano de Servicios Públicos. 2005a. "25 Años Más de 'Agüita Amarilla'—Análisis de Plan Maestro." Guayaquil, Ecuador: Observatorio Ciudadano de Servicios Públicos.

———. 2005b. "Resumen Ejecutivo Proyecto." Guayaquil, Ecuador: Observatorio Ciudadano de Servicios Públicos.

———. 2005c. "Observaciones Constitucionales, Legales, de Derechos de Usuarios y Consumidores del Contrato de Concesión de las Operaciones de Interagua, ECAPAG y del Servicio de Agua Potable y Alcantarillado Sanitario y Pluvial de la Ciudad de Guayaquil." Guayaquil, Ecuador: Observatorio Ciudadano de Servicios Públicos.

———. 2005d. "Propuestas para Mejorar la Calidad y el Acceso al Servicio de Agua Potable y Alcantarillado Sanitario y Pluvial de

Guayaquil." Guayaquil, Ecuador: Observatorio Ciudadano de Servicios Públicos.

———. 2005e. "Argumentaciones y Observaciones Técnicas del Contrato de Concesión y las Operaciones de Interagua y ECAPAG en Relación al Servicio de Agua Potable y Alcantarillado." Guayaquil, Ecuador: Observatorio Ciudadano de Servicios Públicos.

———. 2005f. "Consulta Ciudadana Agua y Alcantarillado Tribunal Ciudadano Electoral: Acta de Escrutino Final." Guayaquil, Ecuador: Observatorio Ciudadano de Servicios Públicos.

Ochoa, F. and M. Prieto. 1995. "Estrategia Integral y Plan de Acción para la Prestación Eficiente de los Servicios de Agua Potable y Alcantarillado en la Ciudad de Guayaquil: Volume 1, Resumen Ejecutivo." Washington, DC: Inter-American Development Bank.

Ochoa, H. and J.A. Collazos. 2004. "La Evaluación del Desempeño de las Empresas Privatizadas en Colombia: ¿Coincide con la Experiencia Internacional?" Estudios Gerenciales 93. Cali, Colombia: Universidad ICESI, Facultad de Ciencias Administrativas y Económicas.

Organisation for Economic Co-Operation and Development (OECD). 2000. *Global Trends in Urban Water Supply and Waste Water Financing and Management: Changing Roles for the Public and Private Sectors.* Paris: OECD.

Organismo Superior de la Inversión en Energía (OSINERG). 2004. *Memoria anual.* Lima, Peru: OSINERG.

Organismo Superior de Inversíon Privada en Telecomunicaciones (OSIPTEL). 1999. "El acceso Universal y la Política de Fitel." *Estudios en Telecomunicaciones* 5. Lima, Peru: OSIPTEL.

———. 2001. "Contratos de Concesión con CPT S.A., ENTEL PERU S.A." Lima, Peru: OSIPTEL, Gerencia de comunicación corporativa y Gerencia Legal.

Pollitt, M. 2004. "Electricity Reform in Argentina: Lessons for Developing Countries." Cambridge Working Papers in Economics CWPE 0449. Cambridge, United Kingdom: Cambridge University.

Pombo, C. and M. Ramírez. 2003. "Privatization in Colombia: A Plant Performance Analysis." Research Network Working Paper R-458.

Washington, DC: Inter-American Development Bank, Research Department.

———. 2005. "Privatization in Colombia: A Plant Performance Analysis. In: A. Chong and F. López-de-Silanes, editors. 2005. *Costs and Benefits of Privatization in Latin America*. Stanford, CA and Washington, DC: Stanford University Press and World Bank Press.

Prasad, N. 2005. "Social Policy, Regulation and Private Sector in Water Supply: How Issues of Equity, Access and Affordability Are Addressed." UNRISD Research Proposal. New York: United Nations Research Institute for Social Development.

Proinversión. Various dates. Contratos de privatización Luz del Sur, Edelnor, EdeChancay, EdeCañete, Electro Sur Medio, Electro Centro, Electro Norte, Electro Norte Medio, Electro Noroeste. Lima, Peru: Proinversión.

Public Citizen. 2003. "Water Privatization Fiascos: Broken Promises and Social Turmoil." Washington, DC: Public Citizen.

Rama, M. 1999. "Efficient Public Sector Downsizing." *World Bank Economic Review* 13(1): 1–22.

Ramakrishnan, U. 2004. "Nutrition and Low Birth Weight: From Research to Practice." *American Journal of Clinical Nutrition* 79(1): 17–21.

Ramamurti, R. 1996. *Privatizing Monopolies: Lessons from the Telecommunications and Transport Sectors in Latin America*. Baltimore, MD: Johns Hopkins University Press.

Ramamurti, R. and V. Raymond, editors. 1991. *Privatization and Control of State Owned Enterprises*. EDI Development Studies. Washington, DC: World Bank.

Rao, S. et al. 2001. "Intake of Micronutrient-Rich Foods in Rural Indian Mothers Is Associated with the Size of Their Babies at Birth: Pune Maternal Nutrition Study." *Journal of Nutrition* 131: 1217–1224.

Renkow, M., D. Hallstrom and D. Karanja. 2003. "Rural Infrastructure, Transactions Costs and Market Participation in Kenya." *Journal of Development Economics* 73(1): 349–367.

Renzetti, S. and D. Dupont. 2003. "Ownership and Performance of Water Utilities." *Greener Management International* 42: 9–19.

Rives Argenal, A. 2004. "Private Sector Participation in Municipal Water Systems in Latin America." Washington, DC: American University. Manuscript.

Röller, L.-H. and L. Waverman. 2001. "Telecommunications Infrastructure and Economic Growth: A Simultaneous Approach." *American Economic Review* 91(4): 909–923.

Roy, A. 1951. "Some Thoughts on the Distribution of Earnings." *Oxford Economic Papers* 3: 135–145.

Rubin, D. 1974. "Estimating Causal Effects to Treatments in Randomized and Nonrandomised Studies." *Journal of Educational Psychology* 66: 688–701.

———. 1977. "Assignment to Treatment Group on the Basis of a Covariate." *Journal of Educational Studies* 2: 1–26.

———. 1979. "Using Multivariate Matched Sampling and Regression Adjustment to Control Bias in Observational Studies." *Journal of the American Statistical Association* 74: 318–328.

Saavedra, J. 2003. "Comment on: 'The Distributive Impact of Privatization in Latin America: Evidence from Four Countries.'" *Economía* 3(2): 225–230.

———. 2004. "Comment on 'Labor Market Adjustment in Chile.'" *Economía* 5(1): 214–216.

Sáez, R. 1992. "An Overview of Privatization in Chile: The Episodes, the Results, and the Lessons." Consultancy Report. Santiago, Chile: CIEPLAN.

Saunders, R., J. Wardford and B. Wellenius. 1994. *Telecommunications and Economic Development*. Baltimore, MD: Johns Hopkins University Press.

Sclar, E. 2000. *You Don't Always Get What You Pay For: The Economics of Privatization*. Ithaca, NY: Cornell University Press.

Shapiro, C. and R. Willig. 1990. "Economic Rationales for the Scope of Privatization." In: E. Suleiman and J. Waterbury, editors. *The Political Economy of Public Sector Reform and Privatization*. London: Westview Press.

Shleifer, A. 1998. "State versus Private Ownership." *Journal of Economic Perspectives* 12: 133–150.

Shleifer, A. and R. Vishny. 1994. "Politicians and Firms." *Quarterly Journal of Economics* 46: 995–1025.

Sinn, H.-W. 1992. "Privatization in East Germany." NBER Working Paper 3998. Cambridge, MA: National Bureau of Economic Research.

Smith, D. et al. 2001. "Livelihoods Diversification in Uganda: Patterns and Determinants of Change across Two Rural Districts." *Food Policy* 26: 421–435.

Stigler, G. 1961. "The Economics of Information." *Journal of Political Economy* 69(3): 213–225.

Stiglitz, J. 1985. "Information and Economic Analysis." *Economic Journal* 95(Supplement): 21–41.

———. 2002. "Information and the Change in the Paradigm in Economics." *American Economic Review* 92(3): 460–501.

Superintendencia de Servicios Públicos Domiciliarios. 2002. "Acueducto, Alcantarillado y Aseo 1998–2001." *Revista Super Cifras* 6. Bogotá: Colombia: Superintendencia de Servicios Públicos Domiciliarios.

———. 2005a. "Balance entre Asignaciones municipals, contribuciones y subsidios aplicados a los servicios de acueducto, alcantarillado y aseo, Colombia 2001–2004 (muestreo sobre ciudades grandes, intermedias y pequenas." Bogota: Colombia: Superintendencia de Servicios Públicos Domiciliarios.

———. 2005b. "Percepciones 10 anos después de la nueva legislación." Encuesta del Centro Nacional de Consultoría. Bogotá, Colombia: Superintendencia de Servicios Públicos Domiciliarios.

Swyngedouw, E. 1995. "The Contradictions of Urban Water Provision: A Study of Guayaquil, Ecuador." *Third World Planning Review* 17(4): 387–405.

———. 1997. "Power, Nature, and the City. The Conquest of Water and the Political Ecology of Urbanization in Guayaquil, Ecuador: 1880–1990." *Environment and Planning A* 29(2): 311–332.

———. 2004. *Social Power and the Urbanization of Water: Flows of Power*. Oxford Geographical and Environmental Studies Series. Oxford, United Kingdom: Oxford University Press.

Tansel, A. 1999. "Workers Displaced Due to Privatization in Turkey: Before versus After Displacement." METU Studies in Development 25. Ankara, Turkey: Middle East Technical University.

Tezna, C.A. and L.E. Amézquita. 2002. "Proyecto de Investigación: Evaluación del Desempeño de las Empresas Privatizadas." Cali, Colombia: Universidad ICESI, Facultad de Ciencias Administrativas y Económicas.

Torero, M. 2001. "Impacts of Privatization in Peru on Firm Performance." Lima, Peru: Grupo de Análisis para el Desarrollo (GRADE). Mimeographed document.

Tschang, T., M. Chuladul and T. Thu Le. 2002. "Scaling-up Information Services for Development." Journal of International Development 14: 129–141.

United Nations Development Programme (UNDP). 2001. "Road Map towards the Implementation of the United Nations Millennium Declaration." Report of the Secretary-General. New York: UNDP.

United Nations Economic and Social Council. 2002. Twenty-ninth session, Agenda Item 3: "The right to water." (Arts. 11 and 12 of the International Covenant on Economic, Social and Cultural Rights.) Geneva, Switzerland: United Nations.

Van de Walle, D. 1996. "Infrastructure and Poverty in Viet Nam." Living Standards Measurement Study Working Paper 121. Washington, DC: World Bank.

Vickers, J. and G. Yarrow. 1988. Privatization: An Economic Analysis. Cambridge, MA: MIT Press.

Wallsten, S.J. 1999. "An Empirical Analysis of Competition, Privatization, and Regulation in Africa and Latin America." Working Paper 2136. Washington, DC: World Bank, Regulation and Competition Policy, Development Research Group.

———. 2002. "An Empirical Analysis of Competition, Privatization and Regulation in Africa and Latin America." Journal of Industrial Economics 49(1): 1–19.

Warner, M. and A. Hefetz. 2004. "Pragmatism over Politics: Alternative Service Delivery in Local Government, 1992–2002." In: The Municipal Year Book 2004. Washington, DC: International City County Management Association.

Wellenius, B., V. Foster and C. Malmberg-Calvo. 2004. "Private Provision of Rural Infrastructure Services: Competing for Subsidies." Washington, DC: World Bank. Manuscript.

World Health Organization (WHO) and United Nations Children's Fund (UNICEF). 2004. *Meeting the MDG: Drinking Water and Sanitation Target, A Mid-Term Assessment of Progress*. Geneva, Switzerland and New York: World Health Organization and United Nations Children's Fund.

World Bank. 1995. *Bureaucrats in Business: The Economics and Politics of Government Ownership*. Washington, DC: Oxford University Press/ World Bank.

———. 2001. *World Development Indicators*. CD-ROM. Washington, DC: World Bank.

———. 2002. "Water—The Essence of Life." *Development News*. May 17. Washington, DC: World Bank.

World Bank Energy Sector Management Programme/United Nations Development Programme. 2001. *Peru: Rural Electrification*. Washington, DC: World Bank.

World Heath Organization (WHO). 2001. "Poverty and Health—Evidence and Action in WHO's European Region." Document EUR/RC52/8. Geneva, Switzerland: WHO Regional Committee for Europe.

Yan, B., X. Chen and M. Roberts. 1997. "Firm Level Evidence on Productivity Differentials, Turnover, and Exports in Taiwanese Manufacturing." NBER Working Paper 6235. Cambridge, MA: National Bureau of Economic Research.

Yan, B., S. Chung and M. Roberts. 1998. "Productivity and the Decision to Export: Micro Evidence from Taiwan and South Korea. NBER Working Paper 6558. Cambridge, MA: National Bureau of Economic Research.

Yepes, G. 1999. "Do Cross-Subsidies Help the Poor to Benefit from Water and Wastewater Services? Lessons from Guayaquil." New York and Washington, DC: United Nations Development Programme and World Bank Water and Sanitation Program.

Contributors

Lorena Alcázar has a Ph.D. in Economics from Washington University in St. Louis, Missouri. She is currently a Senior Researcher of GRADE and a member of collegiate bodies of OSIPTEL.

Felipe Barrera-Osorio is an Economist with the World Bank Group, Washington, D.C. He received his Ph.D. in Economics from the University of Maryland, College Park.

Orazio Bellettini is the Executive Director of the Foundation for the Advance of Reform and Opportunity (FARO), Ecuador. He pursued a master's degree in Public Administration from Harvard University.

Paul Carrillo has been Assistant Professor of Economics at The George Washington University since 2006. He received his Ph.D. in Economics at the University of Virginia.

Alberto Chong holds a Ph.D. degree in Economics from Cornell University. He is currently a Principal Research Economist in the Research Department of the Inter-American Development Bank.

Elizabeth Coombs works as an Assistant Researcher at the Foundation for the Advance of Reform and Opportunity (FARO) in Ecuador.

Virgilio Galdo is a graduate student in the Department of Economics, Syracuse University, and holds an M.A. in Economics from Michigan State University.

Sebastián Galiani received his Ph.D. degree from Oxford and is Associate Professor of Economics at Washington University, Saint Louis, Missouri.

Martín González-Eiras received his Ph.D. in Economics from the Massachusetts Institute of Technology. Since 2000, he has been Assistant Professor at Universidad de San Andrés, Buenos Aires, Argentina.

Martín González-Rozada is Professor at the Business School at Universidad Torcuato Di Tella, Buenos Aires, Argentina. He pursued a Ph.D. in Economics at Boston University.

Eduardo Nakasone is a doctoral student in the Agricultural and Resource Economics Department of the University of Maryland.

Mauricio Olivera received his Ph.D. in Economics at George Washington University. He is an Associate Researcher at Fedesarrollo, Bogotá, Colombia.

Martín A. Rossi pursued his Ph.D. in Economics at Oxford University. He is an Assistant Professor at Universidad de San Andrés, Buenos Aires, Argentina.

Ernesto Schargrodsky is Professor at Universidad Torcuato Di Tella in Buenos Aires, Argentina. He received his Ph.D. in Economics from Harvard University.

Máximo Torero is the Division Director of Markets, Trade, and Institutions at the International Food Policy Research Institute (IFPRI). He received his Ph.D. in Economics from the University of California, Los Angeles.

Index

DATE DUE
